HOMOGENEOUS HYDROGENATION IN ORGANIC CHEMISTRY

HOMOGENEOUS CATALYSIS IN ORGANIC AND INORGANIC CHEMISTRY

VOLUME 1

EDITOR: B. R. JAMES

F. J. McQUILLIN
University of Newcastle upon Tyne, England

HOMOGENEOUS HYDROGENATION IN ORGANIC CHEMISTRY

D. REIDEL PUBLISHING COMPANY
DORDRECHT-HOLLAND / BOSTON-U.S.A.

Library of Congress Cataloging in Publication Data

McQuillin, F. J.
Homogeneous hydrogenation in organic chemistry

(Homogeneous catalysis in organic and inorganic chemistry; v. 1)
Includes bibliographical references and index.
1. Hydrogenation. 2. Catalysts. 3. Hydrocarbons.
I. Title. II. Series.
QD281.H8M26 547'.23 75-37874

ISBN-13: 978-94-010-1880-7 e-ISBN-13: 978-94-010-1878-4
DOI: 10.1007/978-94-010-1878-4

Published by D. Reidel Publishing Company.
P.O. Box 17, Dordrecht, Holland

Sold and distributed in the U.S.A., Canada, and Mexico
by D. Reidel Publishing Company, Inc.
Lincoln Building, 160 Old Derby Street, Hingham,
Mass. 02043, U.S.A.

TABLE OF CONTENTS

PREFACE

Organic chemistry is constantly concerned with effecting reactions at a particular centre in a complex molecule, and if possible with a high and predictable level of stereoselectivity. In the light of much accumulated experience within organic chemistry it is usually possible to assess the likelihood of alternative reaction pathways at least qualitatively. However, well based expectations can be falsified, and the experiments directed to the synthesis of vitamin B_{12} which led to Woodward's recognition of orbital symmetry control in organic chemistry are an instructive example. Our limitations in this respect are very much accentuated in the case of heterogeneous reactions, which present additional problems, and except for very well studied instances, heterogeneous catalysis has remained a relatively empirical area of chemistry. Knowledge in this area has, however, been greatly improved by the development of transition metal complexes which replicate the catalytic properties of the metals, and are effective in a homogeneous reaction system. This development has advanced our understanding of catalysis by making it possible to interpret reactions in strictly molecular terms. In addition, these homogeneously active complexes are frequently more selective than their heterogeneous metallic counterparts either in discriminating between different functional centres in a molecule or in offering better stereoselectivity.

Homogeneous catalysts have now been devised for a number of organic chemical reactions, including hydrogenation, carbonylation, polymerisation, and isomerisation and dismutation of alkenes. The potential, and limitations of these methods within organic chemistry will, however, emerge only by wider application in the laboratory. This text is concerned with homogeneous hydrogenation, and its aim is to make the existing information on homogeneous hydrogenation catalysis more directly available to the practicing chemist. The underlying principles of catalytic hydrogenation are considered, but most emphasis is placed on examples, and on experimental conditions for the use of homogeneous catalysts. For this reason attention has been concentrated on catalytically active complexes which are readily prepared, and for which organic chemical applications have been examined. A number of metal complexes which exhibit activity as hydrogenation catalysts have not been included since they have, as yet, found no distinct area of application. Discussion of these catalysts may be found in reviews, notably in B. R. James's 'Homogeneous Hydrogenation', in the *Chemical Society's Specialist Periodical Reports on Organometallic Chemistry* and in the article by Harmon, Gupta and Brown in *Chemical Reviews* **73**, 21 (1973). Useful articles on hydrogenation by R. Coffey and by A. Andreetta, F. Conti and G. F. Ferrari also appear in *Aspects of Homogeneous Catalysis*, Vol. 1, ed. by R. Ugo.

I am grateful to my colleague Dr N. A. Hughes and to Professor B. R. James for reading the text and for a number of useful suggestions.

CHAPTER I

GENERAL PRINCIPLES

1. The Activation of Hydrogen

Although the reaction: alkene + hydrogen → alkane is thermodynamically allowed, hydrogen is a rather stable molecule not easily susceptible to polarisation, and in consequence reaction with an alkene is not observed in absence of a catalyst. Concerted addition is also symmetry disallowed.

Polarisation of the hydrogen molecule may indeed be observed [1], but only under rather severe circumstances. Thus the equilibrium:

$$Me_3CH + H^+ \rightleftarrows Me_3\overset{+}{C} + H_2$$

has been established for solutions in HF—SbF_5, and examined both by the rate of evolution of hydrogen or of the formation of isobutane, or of ^{2}H-isobutane, from reaction of *t*-butyl cation with hydrogen or deuterium [1]. Similar observations [1a] have been recorded for the reaction of benzene and hydrogen in presence of HF—TaF_5 as the acidic medium. Hydrogenation to give cyclohexane is regarded as resulting from a sequence of protonation and hydride transfer:

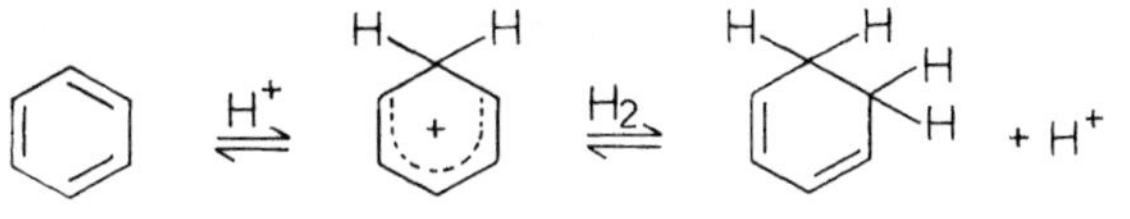

However, although these observations demonstrate polarisation leading to heterolysis of the hydrogen molecule, the reaction requires a very highly acidic medium.

A complementary base induced polarisation of hydrogen has also been realised experimentally [2]. Benzophenone may be reduced by means of hydrogen in presence of potassium *t*-butoxide in *t*-butanol. The reaction sequence may be represented:

$$t\text{-BuO}^- + H_2 \rightleftarrows t\text{-BuOH} + H^-$$
$$Ph_2CO + H^- \rightleftarrows Ph_2CHO^-$$
$$Ph_2CHO^- + t\text{-BuOH} \rightleftarrows Ph_2CHOH + t\text{-BuO}^-.$$

Nitrobenzene has also been reduced by this system. However, the reaction conditions for reduction, viz. 130–230 °C, 75–100 atm pressure of H_2, are again very severe.

By contrast various of the transition metals, such as platinum, palladium, rhodium, nickel, and now also a range of soluble transition metal complexes, are able to catalyse the activation of hydrogen and hydrogenation of olefins or other acceptors at the ordinary temperature and 1 atm hydrogen pressure.

The most essential element in this catalysis is the conversion of hydrogen into a transition metal hydride. In this context, the reversible formation of a dihydride (2) from reaction of hydrogen with *trans*-carbonylchloro *bis* (triphenylphosphine) iridium (I) (1) proved a particularly significant observation [3].

$$\underset{(1)}{trans\ IrCl(CO)(PPh_3)_2} + H_2 \rightarrow \underset{(2)}{IrH_2Cl(CO)(PPh_3)_2} \qquad (i)$$

The dihydride (2) could be characterised [4] by infrared bands due to the metal hydride ligands, ν_{IrH} 2222, 2098 cm^{-1}, and by ^1H n.m.r. hydride signals: τ 28.4, 17.3, each as a 1:2:1 triplet due to ^1H—^{31}P coupling, i.e. consistent with structure (3). Each hydride ligand is *cis* to two phosphorus groups, but the hydrides, which are respectively *trans* to Cl and CO ligands exhibit both different infrared frequencies and different n.m.r. chemical shifts.

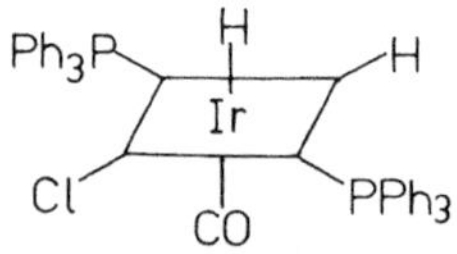

(3)

A further important general point to emerge from studies with carbonyl *bis* (triphenylphosphine) iridium complexes is that as the halogen is varied, the rate of reaction with hydrogen increases [5] viz., Cl < Br < I. Also as the base properties of the phosphine are increased by introducing electron donating para substituents into the phenyl groups, e.g. *p*-MeO or *p*-Me, reaction with hydrogen is facilitated [6]. This emphasises the important influence of the totality of the ligand groups on the process of hydride formation, and on the properties of the hydrido complex which is formed.

In the transition metal hydrides, of which (3) is but one example, the metal hydride bond is covalent and not markedly polar, although in certain carbonyl hydrides such as $FeH_2(CO)_4$, the stability of the metal carbonyl anion, i.e. $[FeH(CO)_4]^-$ or $[Fe(CO)_4]^{2-}$ may render the hydride somewhat acidic [7].

Thermodynamic measurements on reaction (i) [5], and on a parallel instance of transition metal hydride formation shown in reaction (ii), indicate [8] a metal-hydrogen bond energy of ca. 250 kJ mol^{-1}.

$$2[Co(CN)_5]^{3-} + H_2 \rightarrow 2[HCo(CN)_5]^{3-} \qquad \text{(ii)}$$

It is of interest that the value is of much the same order as estimates of the binding energy for hydrogen on the surface of various transition metals [9].

The ability of a catalytic complex to activate hydrogen and form a metal hydride, as in equation (i), is a necessary, but not, in itself, a sufficient condition for catalytic hydrogenation of for example an olefin. For hydrogen transfer to such an acceptor the olefin must also generally be co-ordinated to the metal. Inspection of the structure of the hexa-coordinated dihydride (3) indicates that this co-ordinatively saturated species can accept an olefin ligand only after displacement of one of the ligands already present. The ligands are strongly bound and for this reason carbonylchloro *bis* (triphenylphosphine) iridium (I) (1) under hydrogen is not a particularly effective hydrogenation catalyst [10, 11]. In toluene solution at 70–80° and 1 atm H_2, slow hydrogenation by $IrCl(CO)(PPh_3)_2$ was observed for ethylene, propylene and for other simple olefins, but only over many hours. Nevertheless, this particular hydride-forming complex is both important historically, and also in drawing attention to the dual requirement for an effective catalyst to co-ordinate both hydrogen and the hydrogen acceptor.

From this conclusion it is also evident that the hydrogenation reaction may in principle follow either a sequence: (a) $L_n \cdot$metal hydride + olefin → complex → alkane, known as the hydride route of reaction, or (b): $L_n \cdot$metal + olefin → complex which with hydrogen → hydrido olefin complex → alkane, known as the unsaturate route of reaction (L_n signifies the remaining ligands in the complex).

Kinetic studies [12] with $IrCl(CO)(PPh_3)_2$ have been interpreted in terms of a sequence of type (b), i.e. *via* the unsaturate route:

$IrCl(CO)(PPh_3)_2 + \text{olefin} + \text{solvent} \rightleftarrows IrCl(CO)(PPh_3)$ 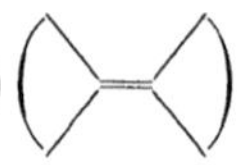 $+ PPh_3$

$\overset{H_2}{\rightleftarrows} IrH_2(Cl)(CO)(PPh_3)$ 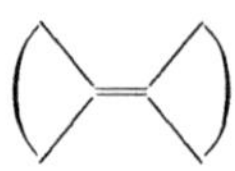 $\rightarrow \text{alkane} + IrCl(CO)(PPh_3)(\text{solvent})$.

This reaction sequence also draws attention to an important role of the solvent in homogeneous catalysis not only in solvating the various components of the system, but in temporarily occupying a co-ordination site at the metal complex, albeit with only weak bonding, and assisting ligand displacement.

This discussion of the behaviour of $IrCl(CO)(PPh_3)_2$ is an appropriate introduction to considering the behaviour of the related rhodium complexes e.g. $RhCl(CO)(PPh_3)_2$ and $RhCl(PPh_3)_3$. Unlike its iridium analogue (1), $RhCl(CO)(PPh_3)_2$ gives no evidence of hydride formation [13] in solution under 1 atm H_2. The $RhCl(PPh_3)_3$ complex, on the other hand, readily forms a dihydrido-derivative in solution, and equally important is a concomitant dissociation to a small extent of a phosphine ligand, viz.:

$$\underset{(4)}{RhCl(PPh_3)_3} + H_2 \rightleftarrows RhH_2Cl(PPh_3)_3 \overset{-PPh_3}{\rightleftarrows} \underset{(5)}{RhH_2Cl(PPh_3)_2} \qquad \text{(iii)}$$

Although reaction (iii) is part of a kinetically complex [14] system which will be discussed more fully later, the immediately important point is that complex (5) is effectively five co-ordinate, and hence has a free site for co-ordination of an olefinic ligand.

The dihydro complex (6) *bis* (triphenylphosphine) dihydridochlororho-

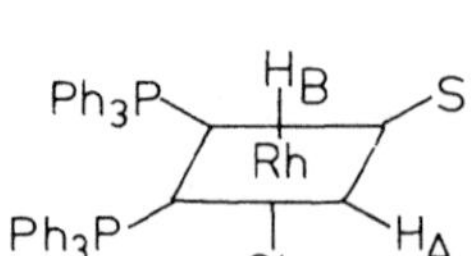

(6)

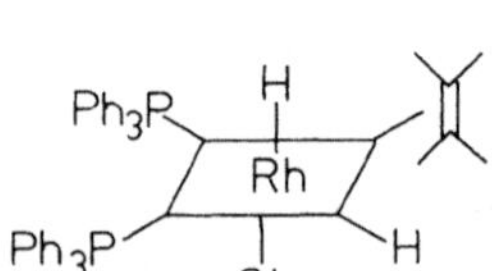

(7)

dium, has been isolated and fully characterised [15]. It shows characteristic ν_{RhH} absorption (2078, 2013 cm^{-1}), δ_{Rh-H} (785 cm^{-1}), and ^{1}H n.m.r. signals: τ 28.2, 21.5, 18.8 which are consistent with structure (6), i.e. with marked ^1H—^{31}P coupling to H_A and small ^1H—^{31}P coupling to H_B. In solution the sixth co-ordination position in (6) is considered to be occupied by a weakly held solvent molecule (S), and a crystalline solvate, $RhH_2Cl(PPh_3)_2 \cdot CH_2Cl_2$ with dichloromethane has indeed been isolated. This easily displaced solvent molecule provides a site for co-ordination of an olefinic acceptor, as in (7). This catalyst has indeed proved very effective for homogeneous hydrogenation [16], viz.:

$$RhCl(PPh_3)_3 + H_2 \xrightleftharpoons{-PPh_3} RhH_2Cl(PPh_3)_2 \rightleftarrows RhH_2$$
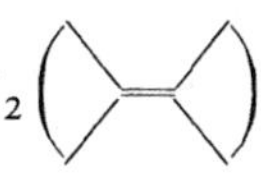
$$Cl(PPh_3)_2 \rightarrow \text{alkane} + RhCl(PPh_3)_2.$$

The reaction sequence reforms $RhCl(PPh_3)_2$ which then re-enters the catalytic cycle. It will be evident, however, that although an intermediate such as (7) is sterically well arranged to permit hydrogen transfer from rhodium to carbon, polarisation of the olefin through co-ordination, as well as a nice balance of Rh—H and C—H bond energies is also necessary to permit a rapid catalytic process of low activation energy.

Chloro*tris*(triphenylphosphine)rhodium (I) (4), which is commonly referred to as Wilkinson's complex, has proved extremely valuable not only as a catalyst, but as a means of clarifying our understanding of catalytic hydrogenation. The iridium analogue, $IrCl(PPh_3)_3$, also reacts readily to form a dihydride:

$$\underset{(8)}{IrCl(PPh_3)_3} + H_2 \rightleftarrows \underset{(9)}{IrH_2Cl(PPh_3)_3}$$

but in this case there is no detectable accompanying phosphine dissociation. The hydride (9) is not active as a catalyst for hydrogenation since there is no readily available vacant site for olefin co-ordination [17]. This interpretation is confirmed by the observation [18] that chloro*bis*(triphenylphosphine)iridium(I) (10) is found to be active for the hydrogenation of alkenes. Reaction may be formulated either by analogy with the rhodium analogue, *via* a dihydride (11), which still retains a free co-ordination site, or possibly through reaction of an olefin co-ordinate (12) with hydrogen.

$$\underset{(10)}{IrCl(PPh_3)_2} + H_2 \rightleftarrows \underset{(11)}{IrH_2Cl(PPh_3)_2}$$

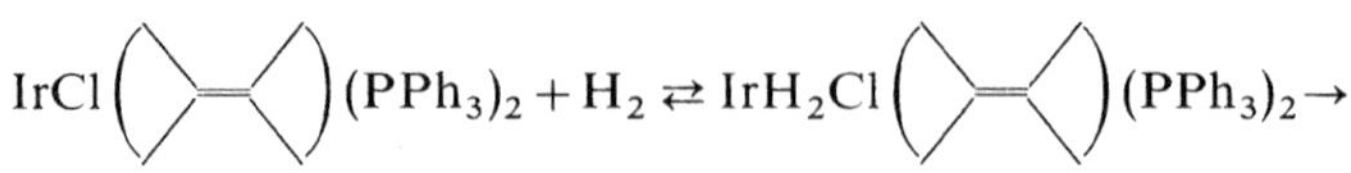

(12) $\rightarrow$alkane + $IrCl(PPh_3)_2$

Ruthenium also yields a catalytically active hydride complex, RuHCl $(PPh_3)_3$, i.e. chlorohydrido*tris*(triphenylphosphine)ruthenium(II) [19]. This differs from the rhodium and iridium complexes (6) and (11) above in that it is a monohydride. There is also a difference in the mode of formation. The dihydrides $RhH_2Cl(PPh_3)_2$ and $IrH_2Cl(PPh_3)_3$ or $IrH_2Cl(CO)(PPh_3)_2$ are formed by direct addition of hydrogen, so that the formal oxidation level of the metal is increased by two units. The ruthenium monohydride RuHCl $(PPh_3)_3$ is obtained by displacement from the dichloride:

$$RuCl_2(PPh_3)_3 + H_2 \rightarrow RuHCl(PPh_3)_3 + HCl$$

The process is assisted by use of a solvent such as ethanol which solvates the hydrogen chloride formed, or by an added base, e.g. triethylamine which removes HCl. Sodium borohydride reduction is, however, a more convenient method of preparation.

From molecular weight [19a] and spectral [19b, 19c] measurements there is evidence for phosphine dissociation from $RuCl_2(PPh_3)_3$ in solution:

$$RuCl_2(PPh_3)_3 \rightleftarrows RuCl_2(PPh_3)_2 + PPh_3.$$

This would leave free, or solvated, two co-ordination sites which permits halogen displacement *via* formation of a dihydride, i.e. following the pattern of hydrogen addition shown by the rhodium and iridium complexes considered above, viz.:

$$RuCl_2(PPh_3)_2 + H_2 \rightleftarrows RuH_2Cl_2(PPh_3)_2$$
$$\xrightarrow{-HCl} RuHCl(PPh_3)_2 \xrightarrow{PPh_3} RuHCl(PPh_3)_3.$$

A series of ruthenium hydrido complexes, $RuH(OCOR)(PPh_3)_3$ containing a carboxylate residue in place of chloride as the anionic ligand have also been described [20], and shown to be effective homogeneous catalysts for hydrogenation. The survey has covered complexes containing anionic ligands —OCOR where R = Me, Et, *n*-Pr, *i*-Pr, CMe_3, Ph, CH_2Cl and CF_3. In the solid state, these carboxylate derivatives appear to be rather less air sensitive than the chloro complex, $RuHCl(PPh_3)_3$, but these complexes as a group are found to be highly oxygen sensitive in solution [20].

A monohydrido-rhodium complex which is also active as a catalyst for homogeneous hydrogenation is carbonylhydrido*tris*(triphenylphosphine) rhodium (I) [21], $RhH(CO)(PPh_3)_3$, which is conveniently available by sodium borohydride reduction of the corresponding chloride:

$$RhCl(CO)(PPh_3)_3 \xrightarrow{NaBH_4} RhH(CO)(PPh_3)_3.$$

This hydride is characterised [21] by ν_{RhH} 2000 cm^{-1}, and by a ^1H n.m.r. hydride signal as a doublet of quartets: τ 19.69, J_{PH} 14, J_{RhH} 1 Hz at $-35°$, which however, broadens at higher temperatures due to phosphine exchange arising from dissociation:

$$RhH(CO)(PPh_3)_3 \rightleftarrows RhH(CO)(PPh_3)_2 + PPh_3.$$

Hence, there is good evidence that in solution carbonylhydrido *tris*(triphenylphosphine)rhodium (I) will provide a species with free co-ordination sites, in principle capable of accepting an olefinic ligand. This is also clearly established by the observed [21] rapid ^1H—^2H exchange of the hydrido complex in benzene solution under deuterium, a process which presumably involves phosphine dissociation, deuterium addition, elimination of HD, and re-coordination of phosphine:

$$\begin{aligned} &RhH(CO)(PPh_3)_3 \rightleftarrows RhH(CO)(PPh_3)_2 + PPh_3 \\ &\overset{D_2}{\rightleftarrows} RhHD_2(CO)(PPh_3)_2 \rightleftarrows RhD(CO)(PPh_3)_2 + \\ &+ HD \xrightarrow{PPh_3} RhD(CO)(PPh_3)_3. \end{aligned}$$

2. Mode of Reaction of Hydrido-Complexes with Alkenes

Although further examples of practically useful homogeneous catalyst systems will be considered below it is convenient at this point to examine the catalytic hydrogenation reaction sequence in terms of the fully characterised hydrido complexes which have already been described.

The most thoroughly investigated example is the dihydride (6) derived from chloro*tris*(triphenylphosphine)rhodium (I). The behaviour of this complex also serves to illustrate the difference between the two possible routes of hydrogenation catalysis, namely (i) formation of a hydride followed by co-ordination of alkene, or (ii) co-ordination of alkene followed by reaction with hydrogen.

Since all the ligands co-ordinated to a metal centre determine the energy levels, and since a hydride and an alkene ligand are not identical in their electronic effects, sequences (i) and (ii) are not necessarily equally facile. This principle is well illustrated by the observation [15] that whereas RhCl $(PPh_3)_3$ reacts rapidly with hydrogen and the derived hydride, RhH_2Cl $(PPh_3)_2$, hydrogenates ethylene, $RhCl(PPh_3)_3$ forms a co-ordinate with ethylene, $RhCl(C_2H_4)(PPh_3)_2$, which it is reported [15] fails to activate hydrogen at 1 atm pressure i.e.:

$$RhH_2Cl(PPh_3)_2 + C_2H_4 \longrightarrow RhCl(PPh_3)_2 + C_2H_6$$
$$RhCl(C_2H_4)(PPh_3)_2 + H_2 \not\longrightarrow C_2H_6.$$

The explanation [15] is considered to lie in the marked π-acid properties of ethylene as a ligand, which reduces the reactivity of the metal centre towards hydrogen. However, formation of the ethylene complex is reversible, and ethylene dissociation will clearly reopen the alternative route of reaction *via* the dihydride, $RhCl(H_2)(PPh_3)_2$. This may explain the observed hydrogenation of ethylene by $RhCl(PPh_3)_3$ under certain circumstances [22], and the catalytic activity of the ethylene complex, $RhCl(C_2H_4)(PPh_3)_2$ in the hydrogenation of other alkenes [23].

A reaction sequence indicating the principal steps for hydrogenation with $RhCl(PPh_3)_3$ may be outlined:

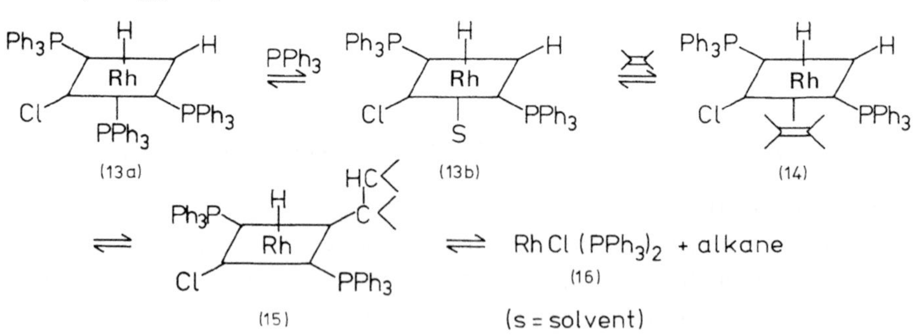

Scheme 1.

There is good evidence, discussed later, that the two hydrogens of the dihydride (13) are transferred sequentially in a two step process, and that stage (14)→(15) leading to the transient rhodium alkyl intermediate is reversible. In (13) the environments of the two hydride ligands are non-equivalent, and in any case, with an unsymmetrically substituted alkene, hydrogen transfer to the two terminii would be expected to occur at different rates. With this rhodium complex, however, stage (15)→(16) is generally very fast, and the intrinsic reversibility of stage (14)→(15), which is considered more fully under the heading of catalysed olefin isomerisation, generally has no important consequences.

The kinetics of hydrogenation using $RhCl(PPh_3)_3$ generally support the reaction sequence outlined above. The rate data are found [15] to fit reasonably the general relation:

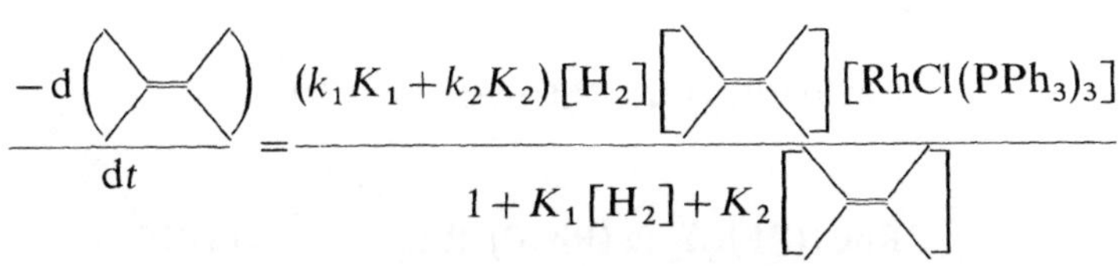

where K_1, K_2 and k_1, k_2 are defined in the reaction scheme:

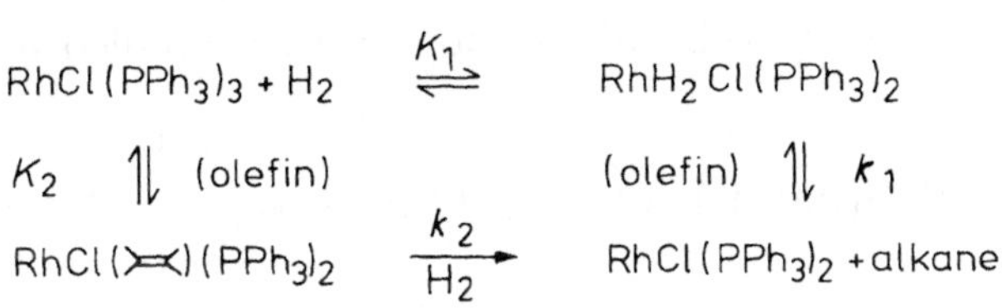

and k_2 is ~ 0, i.e. the olefin complex does not contribute to the reaction.

Close examination, however, has revealed that the behaviour of chloro-*tris*(triphenylphosphine)rhodium (I) in solution is rather complex [22, 24b]. It has been shown that $RhCl(PPh_3)_3$ dissociates in solution:

$$RhCl(PPh_3)_3 \rightleftarrows RhCl(PPh_3)_2 + PPh_3$$

and that although this equilibrium lies far to the left ($K = 1.4 \times 10^{-4}$ M), the dissociated species reacts very rapidly with hydrogen:

$$RhCl(PPh_3)_2 + H_2 \rightleftarrows RhH_2Cl(PPh_3)_2 .$$

In the presence of the phosphine in solution this chlorodihydrido *bis*(triphenylphosphine) rhodium (III) is in equilibrium with the six co-ordinate $RhH_2Cl(PPh_3)_3$:

$$RhH_2Cl(PPh_3)_2 + PPh_3 \rightleftarrows RhH_2Cl(PPh_3)_3$$

which has also been characterised [22].

These reactions are rapid and appear to contribute the principal components concerned in hydrogenation with $RhCl(PPh_3)_3$ in absence of any appreciable excess of phosphine. The complex $RhCl(PPh_3)_2$ reacts with hydrogen some 10^4 times more rapidly than does $RhCl(PPh_3)_3$:

$$RhCl(PPh_3)_3 + H_2 \rightleftarrows RhH_2Cl(PPh_3)_3.$$

Chloro*bis*(triphenylphosphine) rhodium (I) also generates a chloride bridged dimer:

$$RhCl(PPh_3)_3 \rightleftarrows RhCl(PPh_3)_2 + PPh_3$$

$$2RhCl(PPh_3)_2 \rightleftarrows (Ph_3P)_2Rh\langle{}^{Cl}_{Cl}\rangle Rh(PPh_3)_2.$$

This dimer also reacts with hydrogen to give a dihydro-derivative [22].

$$[RhCl(PPh_3)_2]_2 + H_2 \rightleftarrows (PPh_3)_2RhH_2\langle{}^{Cl}_{Cl}\rangle Rh(PPh_3)_2.$$

However, although the hydride from the dimer is catalytically quite effective [22] its rate of formation is slow compared with the rate of formation of $RhH_2Cl(PPh_3)_2$ which is the main reactive species in hydrogenation.

It is evident that these equilibria will be rather sensitive to any excess phosphine which may contaminate a preparation of $RhCl(PPh_3)_3$.

The sequence outlined for catalysis by chloro*tris*(triphenylphosphine) rhodium (I) may be taken as a typical for a catalytically active dihydrido complex, but for a monohydrido catalyst the reaction sequence is necessarily somewhat different. Thus for catalysis of hydrogenation by $RhH(CO)(PPh_3)_3$ (17) the following reaction sequence has been proposed [21] as being in best agreement with the experimental observations.

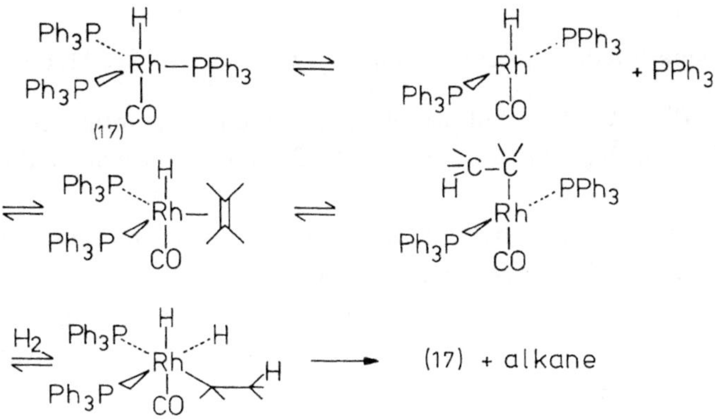

Scheme 2.

Hydrogenation with the mono hydrido catalyst $RuHCl(PPh_3)_3$ may also be accommodated within a reaction sequence similar to Scheme 2. Chlorohydrido *tris*(triphenylphosphine) ruthenium (II) does not complex strongly with terminal alkenes [19a], but in $CDCl_3$ solution with ethylene under pressure (~35 atm) the ^{1}H n.m.r. spectrum viz.: τ 9.0 (triplet) and 7.96 (quartet) gave evidence for formation of an ethyl complex:

$$RuHCl(PPh_3)_3 + C_2H_4 \rightleftarrows RuCl(C_2H_5)(PPh_3)_3$$

accompanied by a colour change from red-violet to brown.

3. The Metal-Alkyl Intermediate

In the reaction sequences outlined in Schemes 1 and 2 there are essentially two different steps in hydrogen transfer to the alkene, namely (i) metal hydride addition across the olefinic bond of the co-ordinated alkene: MH + $>C{=}C<$ → M—C—CH, and (ii) insertion of a hydride ligand into a metal-alkyl bond: H—M—C—CH → M + alkane. With the catalytically very efficient complexes based on rhodium or ruthenium the alkyl intermediate M—C—CH has not been isolated i.e. step (ii) is fast. However, with hydridopentacyanocobaltate (III) ion which has long been known as a catalyst for hydrogenation of conjugated olefins: 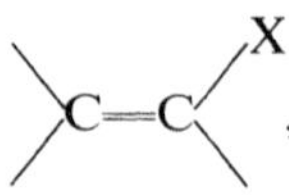 , $X = CH{:}CH_2, CO_2H, CO_2^-$, CN, Ph, α-C_5H_4N, but not of simple alkenes, the intermediate cobalt alkyl derivatives have been isolated [24a, 25a] and characterised in several cases.

Thus, $HCo(CN)_5^{3-}$, as the potassium salt, in aqueous methanol with vinyl cyanide gave a solid adduct analysing correctly as $K_3Co(CN)_5(C_2H_4CN)$ with characteristic ν_{CN} bands and ^1H n.m.r. signals: τ 8.44 (d, J 7 Hz, 3H), and 7.78 (m, 1H), consistent with a structure (18). Thus the hydride adds β to the CN group of $CH_2{=}CHCN$ in a

$$[(CN)_5Co{-}\underset{}{\overset{\displaystyle Me\atop |}{C}}HCN]^{3-}$$

(18)

manner to be expected of nucleophilic hydrogen.

Spectral data have been obtained [24, 25] also for adducts of $HCo(CN)_5^{3-}$ with $PhCOCH{:}CH_2$, 2-vinyl pyridine, and in other cases, but the adducts could not always be isolated in a pure state.

The adduct of $HCo(CN)_5^{3-}$ with butadiene is of particular interest. The isolated solid showed [25a], n.m.r. signals τ 8.26 (d, 3H), 7.44 (m, 2H) and 4.2 (m, 2H), i.e. indicating a structure (19) rather than (20). This conclusion was supported

$$[(CN)_5Co{-}CH_2CH{:}CHCH_3]^{3-} \quad (19)$$

$$[(CN)_5Co\overset{\displaystyle CH_3\atop |}{C}H{-}CH{:}CH_2]^{3-} \quad (20)$$

by the observation that the same complex is obtained from reaction of $Co(CN)_5^{3-}$ with crotyl bromide:

$$CH_3CH{:}CHCH_2Br + 2\,Co(CN)_5^{3-} \rightarrow$$
$$\rightarrow BrCo(CN)_5^{3-} + CH_3CH{:}CHCH_2Co(CN)_5^{3-}.$$

However, allylic systems are known to be very prone to rearrangement, and a 1,2 adduct such as (20) may well be [26] the primary product of addition of $HCo(CN)_5^{3-}$ to 1,3-butadiene.

Better resolved spectra of the 1,3-butadiene adduct support structure (19), but indicate the presence of both *cis*- and *trans*-isomers about the olefinic bond [27]. However, more interesting is the change in the n.m.r. spectrum for solutions of this complex where the cyanide concentration is reduced so that CN/Co < 5. The signals characteristic of (19) are then replaced by (a) τ 8.5 (clean doublet, J 6.2 Hz, 3H), (b) 7.55 (d, J 12.3 Hz, 1H), (c) 6.68 (d, J 7.8 Hz, 1H), (d) 6.23 (dq, 1H), and (e) 4.89 (dt, 1H), consistent with a Co-π-allyl structure (21) with the n.m.r. assignments indicated. Thus in absence of sufficient cyanide ion one ligand site is taken up by co-ordination of the olefinic double bond and formation of a cobalt π-allyl. Clearly, also, by reversible cyanide addition-elimination (21) may interconnect structures (19) and (20).

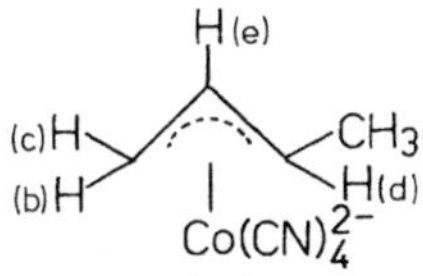

(21)

The π-allyl complex (21) is also important in relation to the experimental observation [28] that the products of reduction of 1,3-dienes in the presence of $HCo(CN)_5^{3-}$ depends on the CN^-/Co ratio as shown in Table I.

TABLE I

Diene	CN^-/Co	Products (%)		
		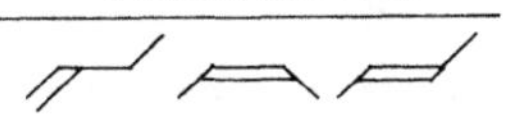		
	5.1	14	6	80
	7.0	92	2	6
		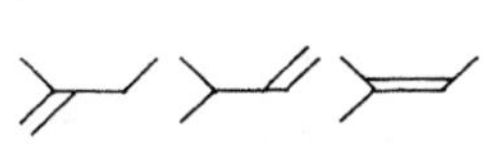		
	5.1	21	1	78
	7.0	91	6	3

It is reasonable that hydrogenolysis of (21) should lead principally to the more stable *trans*-but-2-ene as main product, as is observed under conditions of low CN^-/Co ratio which favour formation of the Co-π-allyl intermediate of type (21). However, to obtain but-1-ene as the main product from complex (19) requires hydrogenolysis by 1,3-displacement, viz.:

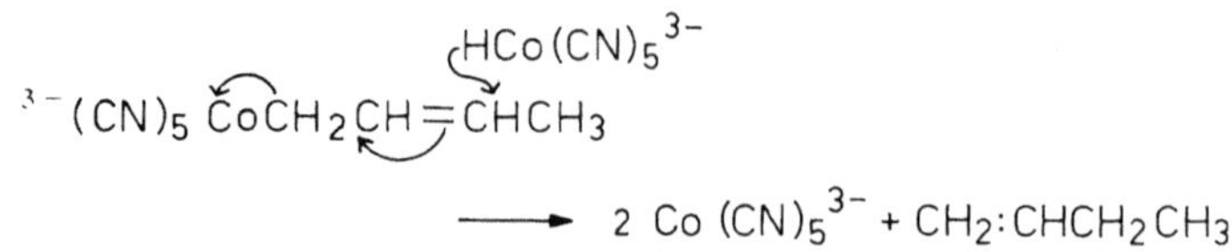

Further data [29] relating to the addition of $HCo(CN)_5^{3-}$ to various unsaturated acids or their anions indicate steric resistance to addition, i.e. $RCH:CHCO_2^-$, R = Me, or COO^-, reacts less rapidly than $CH_2:CRCO_2^-$, R = H, or Me.

A metal alkyl intermediate has also been isolated [30] from addition of $(\pi\text{-}C_5H_5)_2MoH_2$ i.e. cp_2MoH_2 to unsaturated esters such as dimethyl fumarate or maleate, viz.:

$$cp_2MoH_2 + MeO_2CCH:CHCO_2Me \rightarrow$$
$$\rightarrow cp_2Mo(H)[CH(CO_2Me)CH_2CO_2Me].$$

This case is considered more fully later in connection with the stereochemistry of addition of a metal hydride across an olefinic bond. At this point it may, however, be noted that the hydrido-molybdenum alkyl which is formed does not readily decompose to give the alkane derivative, and that cp_2MoH_2 and the related cp_2WH_2 are not very effective catalysts for hydrogenation [31].

4. Homogeneous Hydrogenation Catalysts Without Phosphine, Cyano or Other π-Ligands

In the hydrido complexes so far considered, i.e. $IrCl(CO)(H_2)(PPh_3)_2$, $RhCl(H_2)(PPh_3)_3$, $RuClH(PPh_3)_3$, $HCo(CN)_5^{3-}$, or cp_2MoH_2 the metal hydride is stabilised by the π-acceptor capacity of the ligands PPh_3, CO, CN, cp. Generally in absence of ligands of this type a heavy metal hydride is decomposed to the proton and the metal. However, the electron transfer process appears to be retarded in the presence of certain solvents which apparently act as stabilising ligands in absence of π-acceptors such as PR_3 etc. Maleic acid, for example, could be successfully hydrogenated at 1 atm H_2 by solutions of $RhCl_3$ in dimethylacetamide [32] which acts as a stabilising solvent. The maleic acid itself is also considered to act as a stabilising ligand in a mechanism:

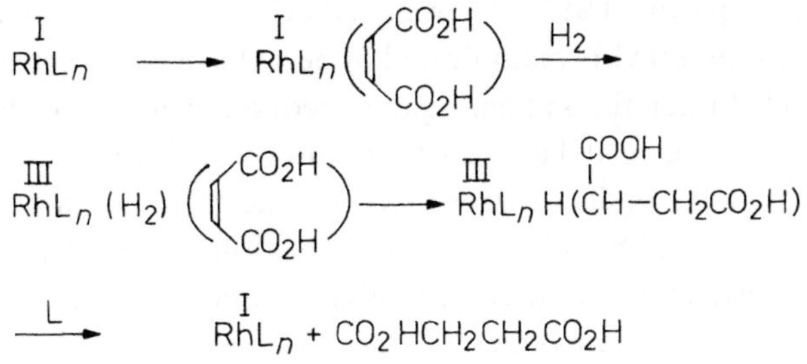

($L = Cl^-$, H_2O, solvent)

It was also possible to reduce fumaric acid and ethylene by this catalyst system the activity of which was shown to depend on initial reduction of rhodium (III) to rhodium (I) by hydrogen [32].

Similar studies have been made using ruthenium chlorides in dimethylacetamide [33], and it has also been shown that $[RuCl_4(bipy)]^{2-}$ activates hydrogen for hydrogenation of maleic acid [34].

These findings indicating that strongly π-acceptor ligands are not essential for metal hydride stability led to the development of effective homogeneous hydrogenation catalysts using Rh(III), or Ni(II) chlorides in dimethylformamide solution.

The trichloro*tris*(pyridine) rhodium complex, $RhCl_3py_3$, in dimethylformamide solution with one equivalent of sodium borohydride was found to give a pink-brown solution which proved very highly active for the hydrogenation of olefinic and other unsaturated systems by hydrogen [35]. The solution proved to be stable to reduction to the metal, and it was possible to isolate a red crystalline active complex $RhCl_2py_2(dmf)(BH_4)$, the composition being established by combustion analysis and ^1H n.m.r. signal intensities. However, it is likely that this complex in solution under hydrogen forms a transient Rh(I) hydrido complex which is the effective catalyst.

Nickel chloride, $NiCl_2 \cdot 6H_2O$, in dimethylformamide with 1 equivalent of sodium borohydride was also found [36] to give a brown solution which catalyses hydrogenation homogeneously and without deposition of nickel. This catalyst was applied particularly to the selective hydrogenation of methyl linoleate, but the catalytically active nickel complex has not been isolated and characterised.

The presence of dimethylformamide as a ligand in the active rhodium complex $RhCl_2py(dmf)BH_4$, as well as other examples, indicates that amides such as dimethylformamide and dimethylacetamide act as protecting ligands, which buffer the system against reduction to the metal by borohydride and hydrogen. It has also been shown [37] that stable solutions may be maintained under hydrogenation conditions when the dimethylformamide is replaced by other solvents to the following extent: diethyleneglycol monoethyl ether or diethylene glycol dimethyl ether 90%, ethylene glycol 65%, water 55%, ethyl alcohol 25%.

Hydrogen activation by solutions of $RhCl_3$ in dimethylsulphoxide has been observed [38], and also by the disulphide complexes, $RhCl_3(R_2S)_3$, $R = Me$ or $PhCH_2$, in dimethylsulphoxide or dimethylacetamide solution. These systems do not, however, appear to offer any particular advantage; strongly co-ordinating olefins such as maleic, fumaric or cinnamic acids may be reduced in the dimethylacetamide medium, but with less strongly co-ordinating olefins e.g. hex-1-ene or cyclohexene the catalyst system is unstable towards hydrogen and ineffective [39].

The tetracarboxylate bridged Rh(II) carboxylates, such as $Rh_2(OCOMe)_4$ are also of interest as examples of catalytically active species which are effective in absence of strongly π-co-ordinating ligands [40].

Dirhodium tetraacetate appears to be most effective in solvents such as dimethyl formamide, dimethyl acetamide, dioxan or tetrahydrofuran. At 25–30° and with 1 atm H_2 these solutions have proved effective for the hydrogenation particularly of terminal alkenes. The catalyst appears to be quite stable in hydrogen and it is not deactivated by oxygen. Kinetic evidence points to the active species being the acetate dimer, $Rh_2(OCOMe)_4$, but spectroscopic data indicate that in the presence of co-ordinating solvents the dimer is terminally solvated as $(S)Rh(O—CMe—O)_4Rh(S)$. It is one of of these terminal positions which is thought to take up hydrogen, but the whole process of catalysis, i.e. hydrogen activation, alkene co-ordination and hydrogen transfer has not been clarified.

A corresponding ruthenium acetate initially formulated [40a] as $Ru_2(OCOMe)_4$ has also been found to catalyse hydrogenation in dimethylformamide solution [40a, 40b] and to be effective with trans disubstituted alkenes and dienes towards which the rhodium acetate catalyst appears inactive. However, the initial formulation has been revised [40c].

In strongly acid solution, e.g. methanolic tetrafluoroboric acid, $Rh_2(OCOMe)_4$ is protonated with consequent loss of acetate ligands and formation of Rh_2^{4+} in solution. This species is not active for hydrogen activation, but becomes active in the presence of triphenylphosphine at 2 mols Ph_3P per atom of rhodium added to the methanol HBF_4 solution [41].

Related ruthenium bridged acetates which have also been studied [41] show similar behaviour on protonation and addition of triphenylphosphine. A range of terminal and in-chain alkenes, dienes and alkynes have been hydrogenated using these catalysts which appear to be rather more reactive than those discussed above, based on the bridged acetates themselves.

5. Catalysts Requiring Higher Pressures and/or Higher Temperatures

It is a characteristic of the series of catalytically active complexes so far described that they are effective in solution under mild conditions, i.e. that any necessary ligand displacement occurs readily with little activation energy and, in particular, that reaction with hydrogen occurs at 1 atm pressure and at the ordinary temperature. These systems are therefore convenient for laboratory work. However, for industrial use where higher pressures and temperatures are more easily handled, various hydride forming metal carbonyls have been examined. Hydrogenation of the aldehyde group: $RCHO \rightarrow RCH_2OH$ is a component reaction in hydroformylation with e.g. $HCo(CO)_4$ as the effective catalyst. This hydridocarbonyl has also been found to effect hydrogenation of for example hept-1-ene, but-1-ene-3-one, acrylonitrile and other unsaturated substances, albeit stoichiometrically [42].

Iron pentacarbonyl with hydrogen at 20 atm and 180° has been shown to hydrogenate linoleate esters [43], and there is reason to believe that the olefinic bonds displace carbon monoxide to give a —$Fe(CO)_n$ type complex which activates hydrogen.

A closely similar catalyst system is found in a series of (arene) $Cr(CO)_3$ complexes which show special value for hydrogenation of 1,3-dienes to *cis*-monoenes [44, 45]. The arene complex is effective for hydrogen deuterium exchange, i.e. for activation of hydrogen [45], but the precise mechanism is not clear. The specific deuteriation of a 1,3-diene:

as well as the formation of a *cis*-monoene product strongly suggests the intervention of a (diene)Cr$(CO)_3$ adduct (22). There is evidence that the arene of the (arene)Cr$(CO)_3$ complex is displaced, and that electron withdrawing groups in the arene assist this process. For this reason a good deal of work has made use of the methyl benzoate complex $(C_6H_5CO_2Me)$Cr $(CO)_3$, but a wide range of other arenes has been investigated, including the particularly effective 1,4-diphenylbuta-1,3-diene derivative.

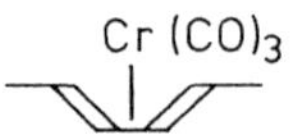

(22)

Some of the dienes successfully hydrogenated by means of $(C_6H_5CO_2Me)$ Cr$(CO)_3$ are given in Table II [46].

TABLE II

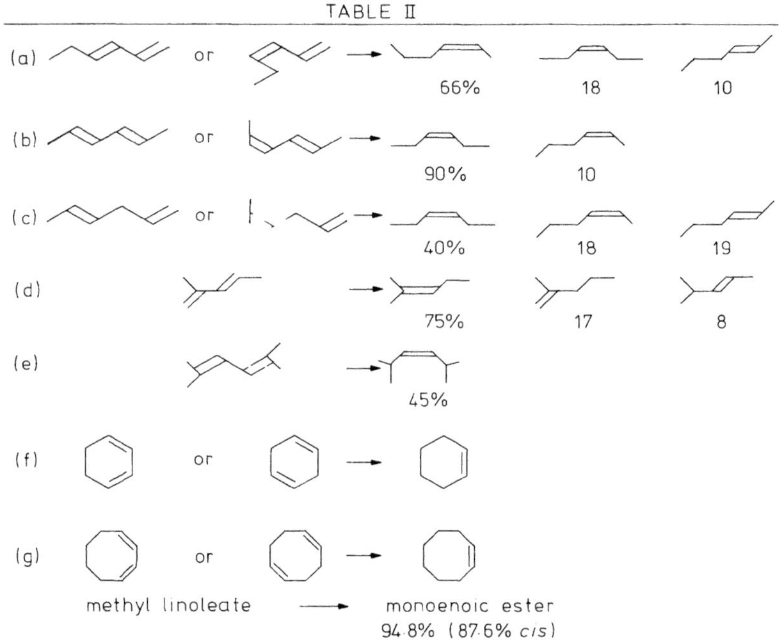

It will be noted from examples (c), (f) and (g) that 1,4- or 1,5- as well as 1,3-dienes may be hydrogenated using this catalyst. There is evidence that unconjugated dienes require prior isomerisation so as to give conjugated species, and that this is effected by this catalyst by hydrogen transfer under the conditions used, i.e. 160 °C and 30 atm H_2.

Inspection of the complex (22) also indicates that 1,4-dialkyl substitution of a 1,3-diene may sterically interfere with complex formation, if the alkyl substituents are *cis*-related. This is found to be the case, and the dimethyl hexadiene in (e), for example, was found to hydrogenate very slowly.

References

1. C. A. F. Bickel, C. J. Gaasbeek, H. Hogeveen, J. M. Oelderik, and J. C. Platteveen: *J. Chem. Soc. Chem. Comm.*, 634 (1967); H. Hogeveen and A. Bickel: *Rec. trav. chim. Pays Bas* **86**, 1313 (1967).

1a. M. Siskin: *J. Amer. Chem. Soc.* **96**, 3641 (1974).

2. C. Walling and L. Bollyky: *J. Amer. Chem. Soc.* **86**, 3750 (1964); cf. **83**, 2968 (1961).
3. L. Vaska and J. W. Diluzio: *J. Amer. Chem. Soc.* **84**, 679 (1962).
4. L. Vaska: *J. Amer. Chem. Soc.* **88**, 4100 (1966); L. Vaska and R. E. Rhodes: *ibid.* **87**, 4970 (1965); R. C. Taylor, J. F. Young, and G. Wilkinson: *Inorg. Chem.* **5**, 20 (1966).
5. L. Vaska and M. F. Werneke: *Trans. New York Acad. Sci.* **31**, 70 (1971).
6. W. Strohmeier and T. Onoda: *Z. Naturforsch.* **23b**, 1927 (1968); **24b**, 515, 931 (1969).
7. Cf. J. Chatt: *Science* **160**, 723 (1968).
8. J. Halpern: *Faraday Soc. Discussion* **46**, 1 (1968).
9. G. C. Bond: *Catalysis by Metals*, Academic Press, London and New York, 1962, p. 84.
10. L. Vaska and R. E. Rhodes: *J. Amer. Chem. Soc.* **87**, 4970 (1965).
11. G. G. Eberhardt and L. Vaska: *J. Catal.* **8**, 183 (1967); B. R. James and N. A. Memon: *Canad. J. Chem.* **46**, 217 (1968); W. Strohmeier and T. Onoda: *Z. Naturforsch.* **24b**, 461, 1493 (1969); W. Strohmeier and R. Fleischmann: *ibid.* **24b**, 1217 (1969); W. Strohmeier, W. Reder-Stirnweiss, and R. Fleischmann: *ibid.* **25b**, 1480 (1970).
12. B. R. James and N. A. Memon: *Canad. J. Chem.* **46**, 217 (1968); M. G. Burnett, R. J. Morrison, and C. J. Strugnell: *J. Chem. Soc. Dalton Trans.* **701** (1973).
13. J. A. Osborn, F. H. Jardine, J. F. Young, and G. Wilkinson: *J. Chem. Soc. A.*, 1711 (1966); L. Vaska and R. E. Rhodes: *J. Amer. Chem. Soc.* **87**, 4970 (1965).
14. J. Halpern and C. S. Wong: *J. Chem. Soc. Chem. Soc.*, 629 (1973).
15. J. A. Osborn, F. H. Jardine, J. F. Young, and G. Wilkinson: *J. Chem. Soc. A*, 1711 (1966).
16. F. H. Jardine, J. A. Osborn, and G. Wilkinson: *J. Chem. Soc. (A)*, 1574 (1967); S. Montelatici, A. van der Ent, J. A. Osborn, and G. Wilkinson: *J. Chem. Soc. (A)*, 1065 (1968).
17. M. A. Bennett and D. L. Milner: *J. Amer. Chem. Soc.* **91**, 6983 (1969).
18. H. van Gaal, H. G. A. M. Cuppers, and A. van der Ent: *J. Chem. Soc. Chem. Comm.*, 1694 (1970).

19a. P. S. Hallman, B. R. McGarvey, and G. Wilkinson: *J. Chem. Soc. (A)*, 3143 (1968).
19b. B. R. James and L. D. Markham: *Inorg. Chem.* **13**, 97 (1974).
19c. K. C. Caulton: *J. Amer. Chem. Soc.* **96**, 3005 (1974).
20. D. Rose, J. D. Gilbert, R. P. Richardson, and G. Wilkinson: *J. Chem. Soc. (A)*, 2610 (1969).
21. D. Evans, G. Yagupsky, and G. Wilkinson: *J. Chem. Soc. (A)*, 2660 (1968); C. O'Connor and G. Wilkinson, *ibid.*, 2665.
22. C. A. Tolman, P. Z. Meakin, D. L. Lindner, and J. P. Jesson: *J. Amer. Chem. Soc.* **96**, 2762 (1974).
23. J. P. Candlin and A. R. Oldham: *Discuss. Faraday Soc.* **46**, 60 (1968).
24a. J. Halpern and J.-Y. Wong: *J. Amer. Chem. Soc.* **90**, 6665 (1968).
24b. H. Arai and J. Halpern: *J. Chem. Soc. Chem. Comm.*, 1571 (1971).
25a. J. Kwiatek and J. K. Seyler: *J. Organometallic Chem.* **3**, 42 (1965).
25b. M. D. Johnson, M. L. Tobe, and L.-Y. Wong: *J. Chem. Soc. (A)*, 929 (1968).
26. M. G. Burnett, P. J. Conolly, and C. Kemball: *J. Chem. Soc. (A)*, 991 (1968).
27. T. Funabiki and K. Tamara: *J. Chem. Soc. Chem. Comm.*, 1177 (1971).
28. J. Kwiatek and J. K. Seyler: *Adv. Chem. Series* **70**, 207 (1968).
29. L. M. Jackman and J. A. Hamilton: *J. Amer. Chem. Soc.* **90**, 1914 (1968).
30. A. Nakamura and S. Otsuka: *J. Amer. Chem. Soc.* **95**, 7262 (1973), *idem.*, *ibid.* **94**, 1886 (1972).
31. Cf. S. Otsuka, A. N kamura, and H. Minamida: *J. Chem. Soc. Chem. Comm.*, 1148 (1969); M. L. H. Green and P. J. Knowles: *ibid.*, 1677 (1970).
32. B. R. James and G. L. Rempel: *Discuss. Faraday Soc.* **46**, 48 (1968).
33. B. R. James: *Homogeneous Hydrogenation*, Wiley-Interscience, 1973, p. 74.
34. B. C. Hui and B. R. James: *Inorg. Nucl. Chem. Letters* **6**, 367 (1970).
35. I. Jardine and F. J. McQuillin: *J. Chem. Soc. Chem. Comm.*, 477 (1969); *J. Chem. Soc. (C)*, 840 (1971).
36. P. Abley and F. J. McQuillin: *J. Catalysis* **24**, 536 (1972); A. G. Hinze and D. J. Frost: *ibid.*, 540.
37. P. Abley: Ph.D. Thesis, University of Newcastle upon Tyne, 1970.
38. B. R. James, F. T. T. Ng, and G. L. Rempel: *Canadian J. Chem.* **47**, 4521 (1969); B. R. James and G. L. Rempel: *Faraday Soc. Discuss.* **46**, 48 (1968).
39. B. R. James and F. T. T. Ng: *J. Chem. Soc. Dalton Trans.*, 355 (1972); S. Montelatici, A. van der Ent, J. A. Osborn, and G. Wilkinson: *J. Chem. Soc. (A)*, 1054 (1968).
40a. B. C. Hui and G. L. Rempel: *J. Chem. Soc. Chem. Comm.*, 1195 (1970).
40b. B. C. Y. Hui, W. K. Yeo, and G. L. Rempel: *Inorg. Chem.* **12**, 757 (1973).
40c. F. A. Cotton, J. G. Norman, A. Spenser, and G. Wilkinson: *J. Chem. Soc. Chem. Comm.*, 967 (1971).
41. P. Legzdins, R. W. Mitchell, G. L. Rempel, J. D. Ruddick, and G. Wilkinson: *J. Chem. Soc. (A)*, 3322 (1970).
42. F. Ungvary and L. Marko: *Acta Chim. Akad. Sci. Hung.* **62**, 425 (1969); R. W. Goetz and M. Orchin: *J. Org. Chem.* **27**, 3698 (1962); *J. Amer. Chem. Soc.* **85**, 2782 (1963).
43. E. N. Frankel, H. M. Peters, E. P. Jones, and H. J. Dutton: *J. Am. Oil Chem. Soc.*, **41**, 186 (1964); cf. *Advances in Chemistry Series* **70**, 177 (1968).

44. M. Cais, E. N. Frankel, and A. Rejoan: *Tetrahedron Letters*, 1919 (1968).
45. E. N. Frankel, E. Selke, and C. A. Glass: *J. Amer. Chem. Soc.* **90**, 2446 (1968).
46. E. N. Frankel and R. O. Butterfield: *J. Org. Chem.* **34**, 3930 (1969); cf. E. N. Frankel, E. Selke, and C. A. Glass: *ibid.*, 3936.

CHAPTER II

THE MECHANISM AND STEREOCHEMISTRY OF HYDROGENATION

1. The Stereochemistry of Hydrogen Addition

Chapter I presented a general outline of the process of catalytic hydrogenation based on a reaction sequence of addition of a metal hydride across an olefinic bond followed by hydride insertion into the carbon-metal σ-bond so formed. The detailed mechanism was not, however, considered.

It has been reasoned [1] that a concerted addition of two hydrogen atoms across an olefinic bond may be an orbital symmetry allowed process in the presence of a transition metal ion, rationalised diagrammatically as in (1) for a metal dihydride MH_2. An analogy has also been drawn with the known process of olefin reduction by diimide, N_2H_2, as shown in (2).

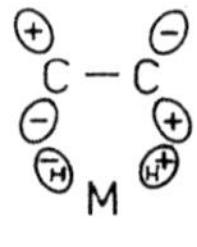

(1)

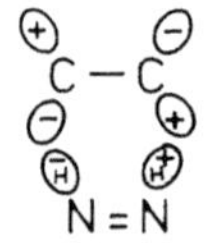

(2)

Both reactions are pictured in terms of overlap of the hydride bond orbitals with the anti-bonding orbital of the olefin. The available evidence, however, indicates that hydrogenation occurs *via* addition of the metal hydride across the double bond of the co-ordinated alkene:

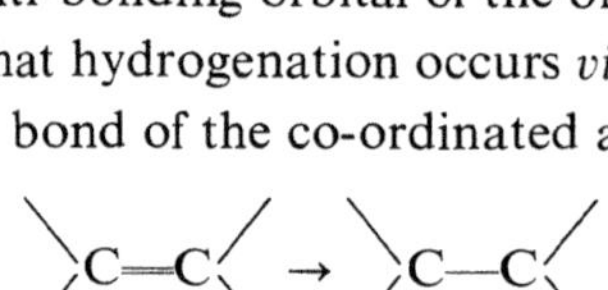

for which the mechanistic picture (1) is inappropriate.

The evidence that this is so comes from the observation that this reaction occurs in a number of cases where the intermediate metal alkyl is sufficiently stable to be observed, i.e. reaction is stepwise, and not a concerted process as is implied in the diagram (1). *Trans*-chlorohydrido*bis*(triethylphosphine) platinum (II) (3) for example reacts with ethylene at 95°, 80 atm pressure, to give the *trans*-chloroethyl complex (4) which may be isolated [2].

$$\underset{(3)}{trans\text{-Pt(H)(Cl)(PEt}_3)_2} \xrightleftharpoons{C_2H_4} \underset{(4)}{trans\text{-PtEt(Cl)(PEt}_3)_2}.$$

This type of process has been observed also for higher alkenes [3].

Undoubtedly the alkyl metal adduct is formed *via* an olefin-metal π-complex, and intermediates of this type have also been observed [4]. Thus, addition of ethylene to *trans*-hydridonitrato*bis*(triethylphosphine)platinum (5) in presence of $NaBPh_4$ gives the cationic ethylene complex (6) in which the hydrido ligand may still be detected by n.m.r. viz. τ 17.2, $J_{^{195}Pt-H}$, 908 Hz, J_{P-H} 12.0 Hz. The cationic nature of the complex arises from the lability of the nitrate ligand.

$$\underset{(5)}{trans\text{-PtH(NO}_3)(\text{PEt}_3)_2} \xrightarrow{C_2H_4} \underset{(6)}{[trans\text{-PtH(C}_2\text{H}_4](\text{PEt}_3)_2]^+ \text{BPh}_4^-}$$

Other hydrido metal olefin complexes of this type are also known [5].

Thus there is experimental evidence for the existence of olefin metal hydride complexes, and for addition of the metal hydride across the alkene bond giving a metal alkyl intermediate. There remains the matter of the stereochemistry of this addition reaction and of the subsequent hydrogenolysis of the metal alkyl to yield the alkane.

With complexes which lead to rapid catalytic hydrogenation it is only possible to determine the overall stereochemistry of addition. This has been demonstrated in many instances, for example:

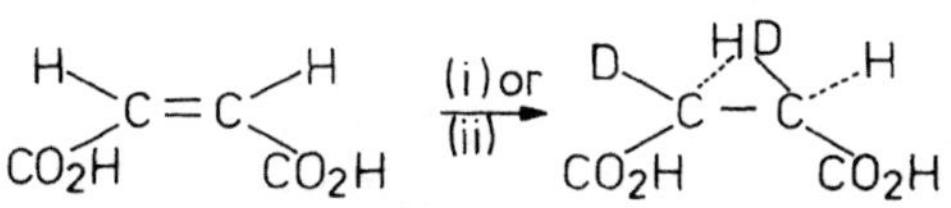

[6,7]

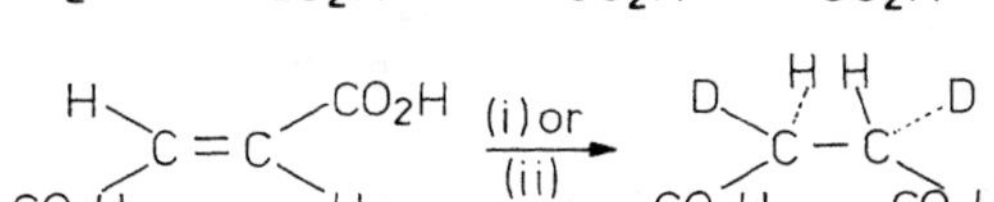

[6.7]

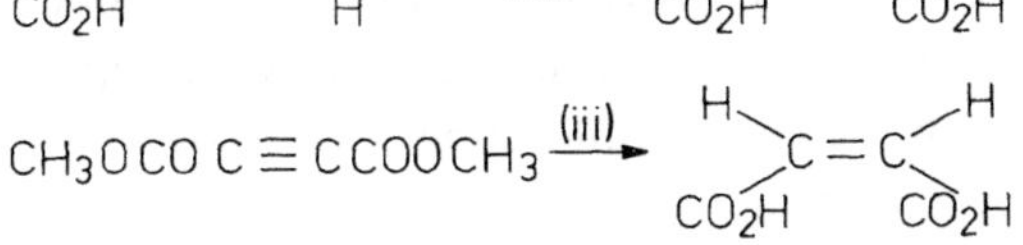

[7]

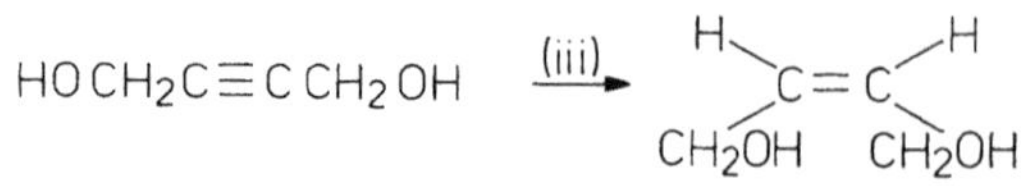

[7]

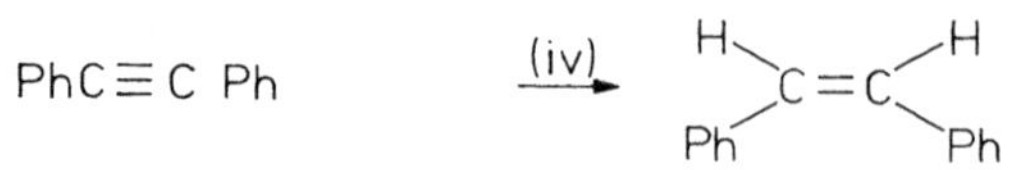

[8]

(i) D_2, $RhCl(PPh_3)_3$ in benzene-ethanol,
(ii) D_2, $RhCl_3py_3$, $NaBD_4$ in dimethylformamide,
(iii) H_2, $RhCl_3py_3$, $NaBH_4$ in dimethylformamide,
(iv) H_2, $RuCl_2(PPh_3)_3$ in benzene.

The implication from these results that metal hydride addition is a *cis* addition process, and that cleavage of the metal alkyl occurs with retention of stereochemistry at carbon is fully substantiated by an instance where the intermediate metal alkyl may be isolated.

In the case of dihydrido*bis*(π-cyclopentadienyl)molybdenum (IV) it is possible to examine the adduct to olefins such as dimethyl maleate or dimethyl fumarate by n.m.r. [9]. Using the corresponding dideuterio complex (7) in deuteriobenzene with dimethyl maleate and dimethyl fumarate led respectively to products in solution which could be characterised as the *cis*-adducts (8) and (9) where $R=CO_2Me$.

cp_2MoD_2
(7)

(8)

(9)

If it may be accepted that the Cp_2MoD moiety will adopt a position of maximal separation from the adjacent carbomethoxy group then the products of *cis* addition, (8) and (9), should show respectively small and large spin-spin coupling constants $J_{H_AH_B}$. In fact the product (8), from dimethyl maleate, shows H_B as a rather broad signal with $J_{H_AH_B}$ 5.5 Hz and J_{H_BD} 2.6 Hz. The product (9) from the fumaric ester, however, shows fairly sharp doublets for both H_A and H_B with $J_{H_AH_B}$ 10.5 Hz.

The n.m.r. data are therefore consistent with structures (8) and (9) arising by cis addition of the molybdenum deuteride complex across the alkene bond.

The $(cp)_2MoH$-alkyl, or $(cp)_2MoD$-alkyl complexes (8) or (9), may be decomposed by heating either to reform the alkene, by reversal of the addition process, or to form the alkane as a result of further hydrogen transfer. Thus reaction of $(cp)_2MoH_2$ with diphenylacetylene led to formation of *cis*-stilbene in solution [9], the released $(cp_2)Mo$ fragment being trapped as the diphenylacetylene complex (10):

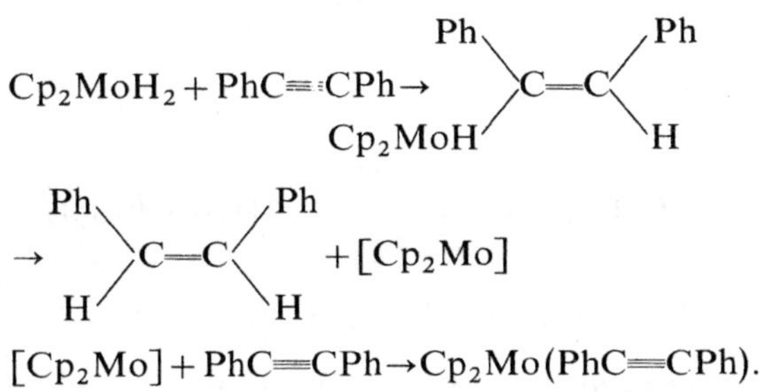
(10)

We may note that when carried out in toluene under reflux (110 °C) this reaction led also to formation of some 20% *trans*-stilbene in the product, indicating a pathway of stereomutation at some stage.

However, the adduct (8) from dimethyl maleate when prepared in solution (benzene) and subsequently heated briefly at 60° gave a product showing infrared bands at 781 and 847 cm^{-1} characteristic of meso-dimethyl 1,2-$^2H_2$-succinate, and the adduct (9) from dimethyl fumarate, treated similarly, gave a product characterised by an infrared band at 812 cm^{-1} attributable to racemic dimethyl 1,2-$^2H_2$-succinate. Thus in both cases there is overall *cis* addition of hydrogen or ^2H.

The results of these studies therefore substantiate the interpretation of homogeneously catalysed hydrogenation in terms of *cis*-addition of the metal hydride across the alkene, or alkyne bond, followed by hydrogen transfer from metal to carbon with retention of carbon stereochemistry.

2. The Reversibility of Metal Hydride Addition to the Alkene Bond

It was indicated above that the addition of *trans*-chlorohydrido*bis*(triethylphosphine)platinum (II) to ethylene is reversible:

$$PtH(Cl)(PEt_3)_2 + C_2H_4 \rightleftarrows PtEt(Cl)(PEt_3)_2.$$

The reverse reaction was studied [3] by synthesis of the deuterium labelled derivative $PtBr(CD_2CH_3)(PEt_3)_2$ (11) *via* the Grignard reaction:

$$trans\text{-}PtBr_2(PEt_3)_2 + CH_3CD_2MgBr \xrightarrow{\text{ether}} PtBr(CD_2CH_3)(PEt_3)_2 \qquad (11)$$

Thermal decomposition of (11) in the solid state *in vacuo* gave a residue showing infrared bands due to Pt—H (2219 cm^{-1}) and Pt—D (1590 cm^{-1}) in a ratio of 52.1 : 23.5 and $^2H_2$- and $^2H_1$-ethylenes in a ratio of 1 : 0.27 were released.

These results are explicable if decomposition takes a pathway:

$$\text{(i)} \qquad PtBr(CD_2CH_3)(PEt_3)_2 \rightleftarrows \begin{matrix} CH_2 \\ \| \\ CD_2 \end{matrix} \text{——} PtH(Br)(PEt_3)_2$$

$$\text{(ii)} \qquad \rightarrow C_2H_nD_{4-n} + PtH(Br)(PEt_3)_2 + PtD(Br)(PEt_3)_2$$

in which step (i) is rapid relative to step (ii), and hydrogen and deuterium can therefore be passed repeatedly between carbon and platinum.

Similar conclusions may be drawn [10] from the results of protolysis of the rhodium complex $[RhCl_2(C_2H_4)_2]^-$ in CH_3OD/DCl which releases ethylene containing some 20% of C_2D_4, viz.:

$$[RhCl_2(C_2H_4)_2]^- + DCl \rightleftarrows [RhDCl_3(C_2H_4)_2]^- \rightleftarrows$$
$$\rightleftarrows [RhCl_3(C_2H_4D)(C_2H_4)]^- \rightleftarrows [RhHCl_3(C_2H_3D)(C_2H_4)]^-$$

etc.

Detailed study [11] of the decomposition of *cis*-$Pt(n\text{-}C_4H_9)_2(PPh_3)_2$ confirms these results. This *bis*(*n*-butyl)platinum derivative is relatively stable, but at 60° in CH_2Cl_2 solution it is decomposed with release of closely equal amounts of but-1-ene and *n*-butane. This is interpreted as the result of a β-elimination process giving a platinum hydride and but-1-ene, followed by hydride insertion into the remaining platinum alkyl bond with release of *n*-butane. There is evidence for an initial phosphine dissociation step, and the overall process may be outlined:

$$Pt(n\text{-}C_4H_9)_2(PPh_3)_2 \rightleftarrows Pt(n\text{-}C_4H_9)_2(PPh_3)+PPh_3 \rightleftarrows$$
$$\rightleftarrows PtH(C_4H_8)(C_4H_9)(PPh_3) \rightleftarrows PtH(C_4H_9)(PPh_3)+C_4H_8 \rightarrow$$
$$\rightarrow Pt(PPh_3)_n+C_4H_{10}.$$

The absence of octane products, i.e. arising by dimerisation of C_4H_9 residues, or of reaction with a radical trap (di-*t*-butyl nitroxyl) and other data exclude a radical decomposition process. Also, evidence for formation of a platinum hydride intermediate was obtained by decomposition of the butyl platinum complex in the presence of hexa-1,5-diene which produced *inter alia* hex-1-ene, and starting from the 2,2-$^2H_2$-butyl complex the hex-1-ene was found to be deuteriated. Further, comparison of the products from decomposition of the ^2H-labelled complexes (12) and (13) showed a closely similar pattern of deuterium labelling in the but-1-ene product, viz.: $^2H_0$ 2%, $^2H_1$ 30%, $^2H_2$ 66%, $^2H_3$ 1%.

$$Pt(CD_2CH_2CH_2CH_3)_2(PPh_3)_2 \quad (12)$$
$$Pt(CH_2CD_2CH_2CH_3)_2(PPh_3)_2 \quad (13)$$

This evidence indicates a rapid and reversible platinum alkyl⇄hydrido- or deuterio-platinum (π-alkene) interchange process.

With alkyl derivatives of metals which are not active in catalytic hydrogen transfer equilibration of this type is not observed. This is seen for example in the decomposition of the copper complex, $(n\text{-}C_4H_9)Cu(PBu_3)$ (at 0° in ether). The mechanism is regarded [12] as a process of β-elimination leading to release of alkene, and subsequent reduction of a second molecule of copper alkyl by the copper hydride which is formed:

$$(n\text{-}C_4H_9)Cu(PBu_3) \rightarrow C_4H_8+HCu(PBu_3)$$
$$(n\text{-}C_4H_9)Cu(PBu_3)+HCu(PBu_3) \rightarrow n\text{-}C_4H_{10}+2Cu+2PBu_3.$$

In agreement with this mechanism the 1,1-$^2H_2$-*n*-butyl copper derivative gave $^2H_2$-but-1-ene, and the 2,2-$^2H_2$-butyl copper complex gave $^2H_1$-but-1-ene together with $^2H_3$-*n*-butane.

These data indicating the reversibility of the process of addition of a metal hydride across an alkene bond are clearly relevant to isomerisation of alkenes during catalytic hydrogenation. A process of addition-elimination may result in trans-location of the olefinic bond, viz.:

$$MH + RCH_2CH{:}CH_2 \rightleftarrows M{-}CH_2CH_2CH_2R +$$
$$+ MCH(CH_3)CH_2R \rightleftarrows MH + RCH{:}CHCH_3.$$

The phenomena also have an evident bearing on the extent to which in catalysed deuteriation an alkene will give mainly the $^2H_2$-alkane, or a mixed deuteriated product. The determining factor will be the relative rates of the addition (i) and hydrogenolysis (ii) steps in hydrogenation:

(i) $MH + RCH{:}CH_2 \rightleftarrows M(CH_2CH_2R) + MCH(CH_3)R$

(ii) $M(CH_2CH_2R)$ or $MCH(CH_3)R + H_2 \rightarrow MH + \text{alkane}.$

For effective catalysis without alkene isomerisation or deuterium scrambling, the irreversible step (ii) should be fast relative to step (i). Whether this is so for any particular catalyst will clearly depend upon the balance of bond energies, and can at present be determined only by experiment.

The experimental data relating to olefin isomerisation and deuteriation are collected in a following chapter.

It may also be noted that although the hydrogenolysis step (ii) above is effectively irreversible under normal conditions of catalysis, it is possible in suitable systems to realise activation of the carbon hydrogen bonds of alkanes. These phenomena are discussed in Chapter X.

References

1. R. G. Pearson: *Accounts Chem. Res.* **4**, 152 (1971).
2. J. Chatt and B. L. Shaw: *J. Chem. Soc.* 5076 (1962); cf. H. C. Clark and C. S. Wong: *J. Amer. Chem. Soc.* **96**, 7213 (1974); H. C. Clark and C. Jablonski: *Inorg. Chem.* **13**, 2213 (1974).
3. J. Chatt, R. S. Coffey, A. Gough, and D. T. Thompson: *J. Chem. Soc. (A)*, 190 (1968).
4. A. J. Deeming, B. F. G. Johnson, and J. Lewis: *J. Chem. Soc. Chem. Comm.*, 598 (1970).
5. M. S. Frazer and W. H. Baddley: *J. Organometal. Chem.* **36**, 377 (1972); R. W. Hazeldine, R. J. Lunt, and R. V. Parish: *J. Chem. Soc. (A)*, 3711 (1971).
6. J. A. Osborn, F. H. Jardine, J. F. Young, and G. Wilkinson: *J. Chem. Soc. A*, 1711 (1966).
7. P. Abley and F. J. McQuillin: *J. Chem. Soc. Chem. Comm.*, 1503 (1969).
8. I. Jardine and F. J. McQuillin: *Tetrahedron Letters*, 4871 (1966).

9. A. Nakamura and S. Otsuka: *J. Amer. Chem. Soc.* **95**, 7262 (1973); *ibid.* **95**, 5091 (1973) and **94**, 1886 (1972).
10. R. Cramer: *Accounts Chem. Res.* **1**, 186 (1968).
11. G. M. Whitesides, J. F. Gaasch, and E. R. Stedronsky: *J. Amer. Chem. Soc.* **94**, 5258 (1972).
12. G. M. Whitesides, E. R. Stedronsky, C. P. Casey, and J. San Filippo: *J. Amer. Chem. Soc.* **92**, 1426 (1970).

CHAPTER III

HOMOGENEOUS HYDROGENATION OF ALKENES, ALKYNES, ALLENES, AND 1,3-DIENES

1. General

Except perhaps with $[HCo(CN)_5]^{3-}$, it is clear that for hydrogenation an alkene or alkyne must first be co-ordinated to the metal centre of the catalytic complex. It is a familiar observation [1] in heterogeneous hydrogenation that alkynes are generally more strongly co-ordinated than alkenes and it is also found [1] that, for an alkyne, alkyl substitution on or near the unsaturated centre may increase, and for an alkene reduce the rate of hydrogenation. It is to be expected that alkyl substitution will in both cases sterically impede absorption, but since reaction depends on activation of hydrogen by the metal-alkyne or metal-alkene complex, too high a surface concentration of a strongly absorbed alkyne molecule is inhibitory, whilst too low a surface concentration is rate limiting.

Somewhat similar considerations apply to homogeneous hydrogenation, i.e. a too stable alkyne or alkene complex may fail to undergo hydrogen addition and heavy substitution may introduce a steric (and electronic) difficulty in alkene or alkyne co-ordination. Homogeneously catalysed hydrogenation has, however, the advantage that discrimination between different structural types in respect of rate of hydrogenation is often more marked than in the heterogeneous process.

It was shown [2] that chloro*tris*(triphenylphosphine) rhodium (I) co-ordinates readily with ethylene to form a complex, $RhCl(C_2H_4)(PPh_3)_2$ which it is reported fails to activate hydrogen. However, complexing between $RhCl(PPh_3)_3$ and propene or hex-1-ene is smaller [2] by a factor of at least 10^3, and co-ordination to 1,2-dialkyl alkenes is further much reduced. Although in terms of rate of hydrogenation these differences prove to be much less marked, alkyl substitution has a large effect. It is to be expected therefore that terminal alkenes should hydrogenate more rapidly than in-chain alkenes, and for the latter that the *cis* isomer should react the more rapidly since of a *cis*, *trans* pair the former is commonly the more easily co-ordinated. Equally, however, strongly co-ordinated alkenes, and especially chelating dienes may react more slowly.

This broad conclusion is reflected in the available data. Thus using hex-1-yne and oct-1-ene with $RhCl(PPh_3)_3$ in a wide range of solvents the alkyne was found [3] to hydrogenate the more rapidly by a factor of ca. 2:1.

The same study [3], however, also drew attention to the effect of electron withdrawing groups in assisting hydrogenation of an alkene. Competitive experiments revealed an order of ease of hydrogenation for substituted alkenes $CH_2{=}CHR$ of $R = CN > CO_2Me > Ph >$ alkyl. The expected effect of alkene stereochemistry was also observed, viz.: diethyl maleate hydrogenated faster than diethyl fumarate, and *cis*-oct-2-ene faster than *trans*-oct-2-ene, and similarly in other examples [4].

Increasing substitution at an alkene centre is, however, of considerable influence, e.g. in the cyclohexene series the following data [5] indicate relative rates of hydrogenation at $RhCl(PPh_3)_3$ in benzene-ethanol:

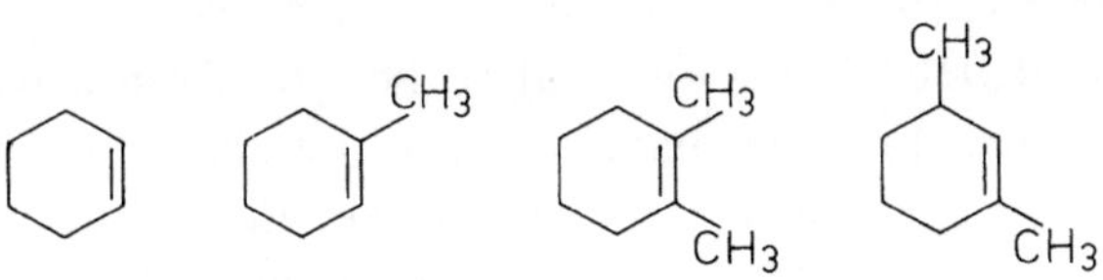

17 0.49 0.0 0.0

Close examination [6] of the hydrogenation of a series of alkenes using $RhCl(PPh_3)_3$ as catalyst in benzene reveals that the effect of increasing substitution is exerted on the entropy of activation, which becomes increasingly negative, and on the enthalpy term which falls. The following data are illustrative:

	$\Delta H^{\ddagger}$ (kcal mole^{-1})	$\Delta S^{\ddagger}$ (e.u.)
$CH_3(CH_2)_3CH{:}CH_2$	18.6	1.1
$(CH_3)_2CH(H)C{=}C(H)CH_3$ (cis)	16.1	−9.3
$(CH_3)_2CH(H)C{=}C(H)CH_3$ (trans)	11.6	−27.6

Thus increasing substitution impedes alkene hydrogenation by imposing limitations on the orientations of the alkene which can lead to effective co-ordination to the metal centre, i.e. $\Delta S^{\ddagger}$ becomes increasingly negative, whilst increased electron density makes the alkene a better electron donor and $\Delta H^{\ddagger}$ falls.

A catalytic complex which has been found [7] to show particularly marked discrimination in favour of the less substituted alkene is chlorohydrido*tris* (triphenylphosphine) ruthenium (II) which effects extremely rapid hydrogenation of terminal alkenes, but only very slow hydrogenation of disubstituted alkenes. Thus, whilst a group of alkenes, $CH_3(CH_2)_nCH{:}CH_2$, $n=2$, 3, 4, 7 are hydrogenated rapidly in the presence of $RuHCl(PPh_3)_3$ in benzene solution, *cis*- and *trans*-$CH_3(CH_2)_nCH{:}CHCH_3$, $n=2,3$, $CH_3(CH_2)_2CH{:}CHCH_2CH_3$, $CH_3CH_2CH_2C(CH_3){:}CH_2$ and cyclohexene react at only 10^{-4} to 10^{-3} of the rate for the first group.

For cycloalkenes, which have been examined [3, 6], cyclopentene, cyclohexene, cycloheptene, cyclo-octene, are all effectively hydrogenated using $RhCl(PPh_3)_3$ in benzene, but at a somewhat decreasing rate with increasing ring size. Using $RuHCl(PPh_3)_3$ as catalyst the cycloalkenes are hydrogenated at only the very slow rate characteristic of 1,2-dialkyl alkenes [7].

Dienes which are conjugated or capable of chelation are frequently found to hydrogenate relatively slowly. Thus norbornadiene (1), cyclo-octadiene (2), and cyclopentadiene (3) are described [3] as being only very slowly hydrogenated using $RhCl(PPh_3)_3$ in benzene. Isoprene (4) and penta-1,3-diene (5) are also found [3, 6] to react slowly using this catalyst, but the non-chelating cyclo-octa-1,3-diene (6) reacts at about the same rate as cyclohexene using $RhCl(PPh_3)_3$ in benzene.

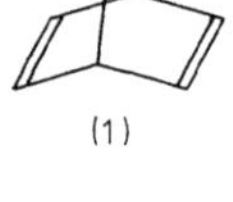
(1)

(2)

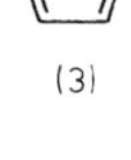
(3)

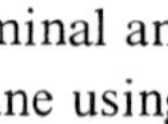
(4)

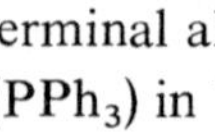
(5)

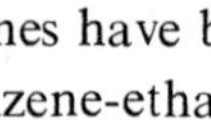
(6)

A number of terminal and non terminal alkynes have been hydrogenated [1, 3, 8] to the alkane using $RhCl(PPh_3)$ in benzene-ethanol. However, terminal alkynes are found to form a stable complex with the ruthenium catalyst, $RuHCl(PPh_3)_3$, and hydrogenation does not occur [7]. Although the rhodium carbonyl hydride complex $RhH(CO)(PPh_3)_3$ has also been reported [9] to react with a terminal acetylene such as hex-1-yne, this alkyne and phenylacetylene have been reduced satisfactorily using this catalyst in toluene, and also, albeit more slowly, with the iridium analogue $IrH(CO)(PPh_3)_3$ as catalyst [10].

The catalyst generated *in situ* from trichloro*tris*(pyridine) rhodium (III), $RhCl_3py_3$ and sodium borohydride in dimethylformamide [11] is found to hydrogenate a wide range of alkenes and alkynes. Terminal alkenes are hydrogenated very rapidly, cycloalkenes show a rate order: cyclohexene > cyclopentene > cycloheptene > cyclo-octene and hydrogenation of norbornene is very rapid. A series of alkynes, e.g., dimethyl acetylenedicarboxylate, or butyn-1,4-diol were also hydrogenated and it was possible to isolate the cis-alkene product after uptake of one equivalent of hydrogen [12].

Hydrogenation carried out in the presence of the catalysts $RhCl(PPh_3)_3$, $RhH(CO)(PPh_3)_3$, $RuHCl(PPh_3)_3$ or $RhCl_3py_3$—$NaBH_4$ does not in general extend beyond reaction of alkene or alkyne centres. Aromatic nuclei are inert, and indeed benzene is a useful solvent in hydrogenation with $RhCl(PPh_3)_3$, and similar complexes. The $RhCl_3py_3$—$NaBH_4$-system has been shown to hydrogenate some hetero aromatics, and certain other groupings, albeit relatively slowly. However, reactive halides, e.g. allyl- or, benzyl halides or α-halo ketones or esters undergo hydrogenolysis with the $RhCl_3py_3/NaBH_4$ catalyst. Except with particular catalyst systems which are discussed later, the carbonyl group is not reduced and the behaviour of carbonyl, halide, nitro, and other groupings towards hydrogenolysis is considered separately in Chapter VI under the heading of the different functional groups.

Experimental details for particular examples of homogeneously catalysed hydrogenation are collected in Chapter XI, but it is appropriate to refer to the order of rate of hydrogen uptake for some typical alkenes or alkynes at this point.

For hept-1-ene at 1 molar concentration in benzene with 1.25 mM $RhCl(PPh_3)_3$ the hydrogen uptake [2] was of the order of 15 ml min^{-1} at 25° which is as high as could be achieved using an active metal catalyst. Rates of uptake of the same order were observed [9] for a range of terminal alkenes at 0.6 M concentration with 1.25 mM $RhH(CO)(PPh_3)_3$. For terminal alkenes with $RuHCl(PPh_3)_3$ rates of ca. 10 times this order may be observed [7].

2. The Hydrogenation of Alkenes

The application of homogeneous hydrogenation catalysis within organic chemistry may be illustrated by a considerable number of instances of the use of chloro*tris*(triphenylphosphine) rhodium (I). The examples also emphasise circumstances where the homogeneous catalyst offers the particular advantage of effecting hydrogenation of less substituted alkenes in the presence of more heavily substituted alkene centres.

This principle is well illustrated by comparing the behaviour of linalool (7) with geraniol (9), or nerol, the 2,3-*cis* isomer of (9). Whereas [13] linalool

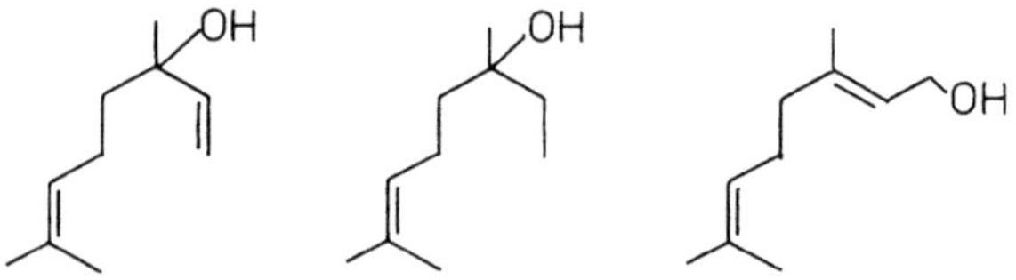

(7) (8) (9)

with $RhCl(PPh_3)$ in benzene is quite rapidly hydrogenated to give the 1,2-dihydro-derivative (8), which proved to be essentially pure to n.m.r. analysis, geraniol and nerol, or their acetates, were found to take up hydrogen only very slowly, and simultaneously at both olefinic centres. In linalool (7) the alkene bonds are respectively mono- and tri-alkyl substituted whereas in geraniol (9) or its 2,3-*cis* isomer, nerol, both alkene centres are tri-substituted.

Similar instances of the same kind are provided by the hydrogenation of carvone (10) to the *p*-menth-1-en-6-one [13] (11), and of (+)-2-methyl-*p*-

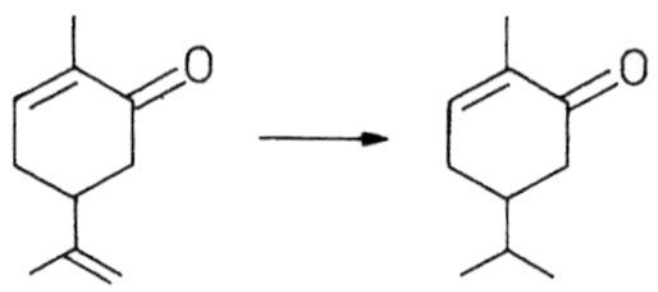

(10) (11)

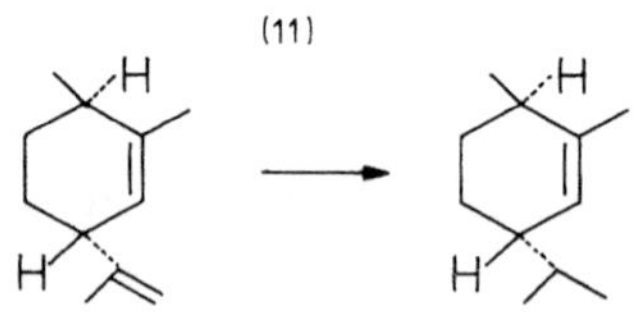

(12) (13)

mentha-2,8-diene (12) to (13). The latter example illustrates the use of homogeneous hydrogenation catalysis in the course of a synthesis of α- and β-cubebene [14].

These examples clearly illustrate the effect of increasing substitution in retarding hydrogenation using $RhCl(PPh_3)_3$ as catalyst. The same principle has also been applied to effect selective hydrogenation in the case of more complex molecules. Eremophilone (14) for example gave [15] the dihydro-derivative (15), and γ-gurjunene [16] (16) the dihydro-derivative (17). Hydrogenation of the isomeric sesquiterpene types (18) and (19) [17] are further examples of the same kind which illustrate selective hydrogenation of a less substituted alkene centre, and in these examples also without reduction of either a carbonyl or a cyclopropane constituent.

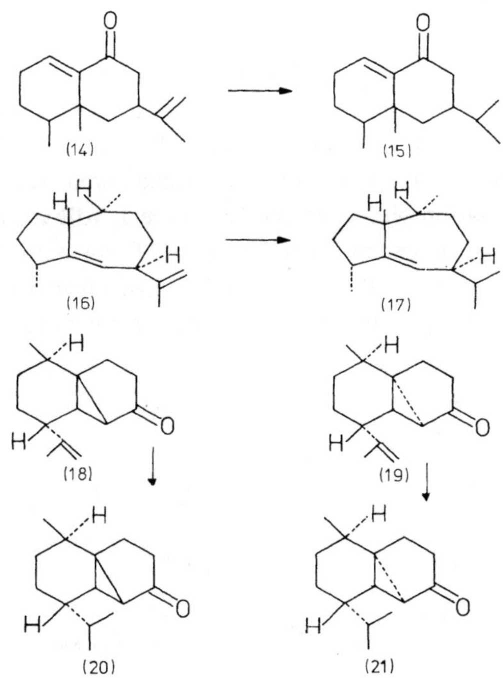

It may also be noted that in the examples (10) and (14) the conjugated carbonyl function does not react either at the carbonyl or alkene centre.

The last two examples (18) and (19) are further illustrations of the use of homogeneous hydrogenation in connexion with sesquiterpene synthesis where it has proved particularly valuable. Further instances of applications in synthesis are to be found in the selective hydrogenation of (+)-nootkatone (22) to the 11,12-dihydro-derivative [18] (23), and of the transformation of (24) to (±)-longicamphenilone (25) the latter in the course of a synthesis of longifolene [19].

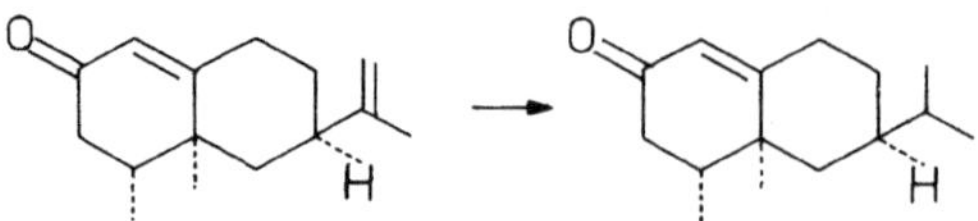

(22) (23)

(24) (25)

It may be noted that the example (24)→(25) does not illustrate selective hydrogenation, but it does exemplify the hydrogenation of a disubstituted alkene bond. However, further instances which illustrate selective hydrogenation of a disubstituted as against a trisubstituted alkene centre are provided by the hydrogenation of α-santonin (26) to the 1,2-dihydro derivative [20] (27), and the analogous example [21] of (28)→(29), where the 4,5-olefinic centre is less heavily substituted than in (26). It is also clear from these examples that alkene conjugation with a carbonyl group does not impede hydrogenation.

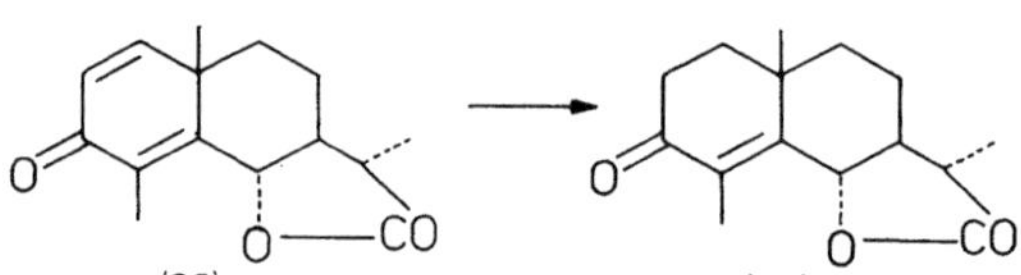

(26) (27)

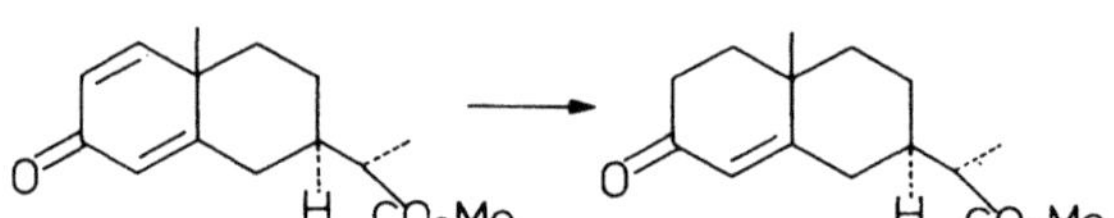

(28) (29)

3. The Influence of Hetero-Atoms

In the above examples to which selective hydrogenation in the presence of $RhCl(PPh_3)_3$ has been successfully applied the olefinic centres differ in the degree of alkyl substitution. An example of a rather different kind is provided by the hydrogenation, using $RhCl(PPh_3)_3$ in benzene, of (−)-dehydrogriseofulvin (30) where the olefinic centres differ not in the degree but in the chemical nature of the substitution [22].

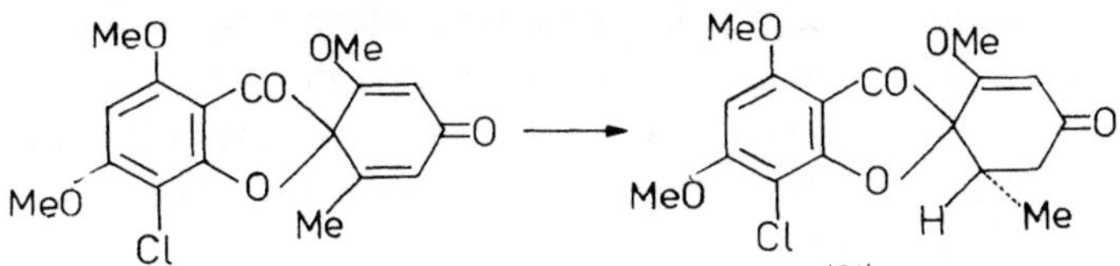

(30) (31)

The formation of griseofulvin (31), in fair yield, indicates the resistance to hydrogenation of the MeOC═C—C═O grouping, which is a vinylogous ester, in comparison with the enone moiety, MeC═C—C═O. However, it may be noted that hydrogen uptake was slow, as is to be expected for a trisubstituted alkene, and some dihydrogriseofulvin was also formed. Nevertheless hydrogenation under these conditions appears to be highly stereoselective; there was no evidence of formation of the isomeric epi-griseofulvin, which is the case in heterogeneous hydrogenation of dehydrogriseofulvin, nor of hydrogenolysis, cf. Chapter XIII.

A further example [22] where heteroatom substitution is decisive is provided by the hydrogenation of the dihydropyranyl ether of cyclohexa-1,4-diene (32), which with $RhCl(PPh_3)_3$ in benzene gave the dihydro derivative (33) the vinyl ether group being unaffected. However, although this vinyl ether system is apparently stable to hydrogenolysis under these conditions,

(32) (33)

such molecules are rather sensitive to acid catalysed hydrolysis. In the analogous case of the methyl ether (34) the hydrogenation product was found to contain cyclohexanone as well as the cyclohexenyl methyl ether

(34) (35)

(35) after hydrogenation using $RhCl(PPh_3)_3$ in benzene, presumably by some acid or Lewis acid assisted hydrolysis. However, in this example also the alkyl ether substituent is effective in impeding hydrogenation of the vinyl ether alkene bond.

In an analogous case (36), where both olefinic bonds are trisubstituted, the product of hydrogenation using $RhCl(PPh_3)_3$ in benzene is reported [22] to be (37), i.e. a methoxy substituent is, as in the case of (30) more inhibitory than a methyl group. The hydrogenation, however, took several days to reach completion.

(36) (37)

An interesting result was encountered [22] in the hydrogenation of the quinones (38) and (39) using the $RhCl(PPh_3)_3$ catalyst. Hydrogen addition was found to take place at the olefinic centre rather than at the carbonyl groups to give the cyclohexendiones rather than the hydroquinones which might have been expected. This principle was applied [22a] to the conversion of the natural product diospyrin (40) to its dihydro derivative (41).

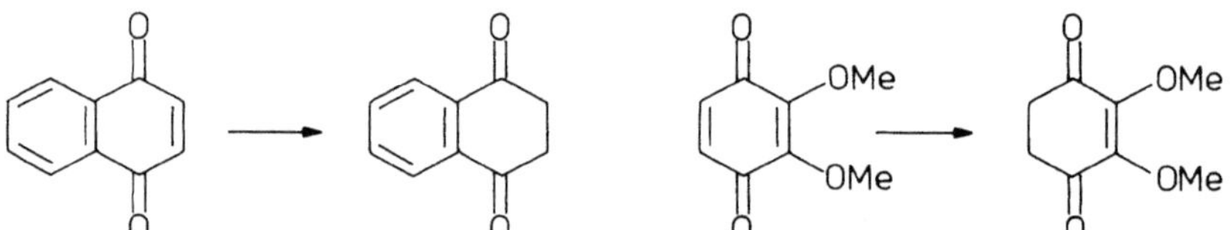

(38) (39)

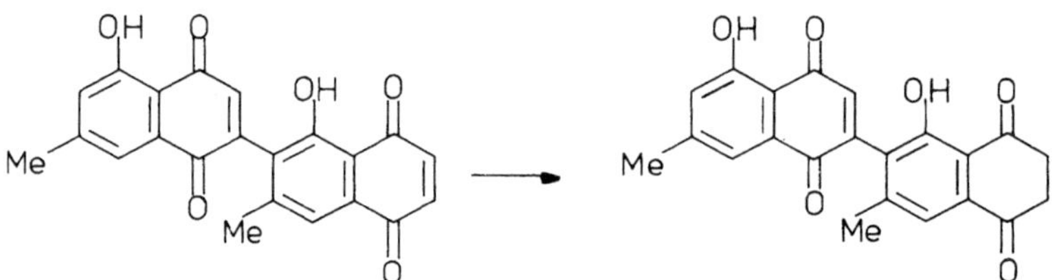

(40) (41)

It is evident from these examples that the presence of a heteroatom substituent has no profound effect on the hydrogenation of an alkene centre. As further instances it may be noted that dihydropyran (42) and the isomeric dimethoxy dihydrofuran derivatives [23] (43) and (44) have been successfully hydrogenated to the dihydro derivatives using $RhCl(PPh_3)_3$.

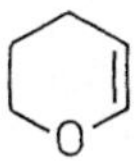

(42)

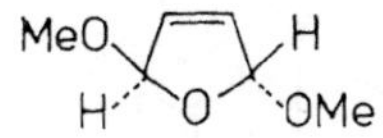

(43)

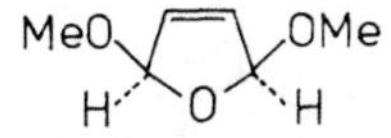

(44)

Using the same catalyst it has also been possible [24] to hydrogenate olefinic centres in a range of thiophen derivatives (45)–(48), i.e. whatever co-ordination may occur through sulphur does not inhibit hydrogenation using $RhCl(PPh_3)_3$.

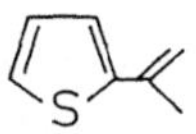

(45)

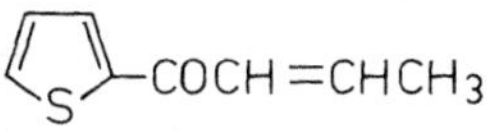

(46)

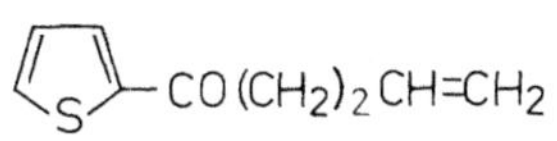

(47)

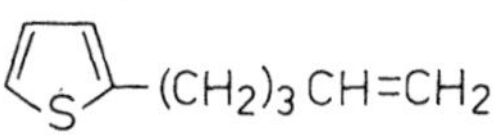

(48)

As an example of a nitrogen containing system reference may be made to the very successful hydrogenation [13] of thebaine (49) to the 8,14-dihydro derivative (50) using $RhCl(PPh_3)_3$ in benzene. Heterogeneous hydrogenation of thebaine is less selective in giving tetrahydro- as well as dihydrothebaine.

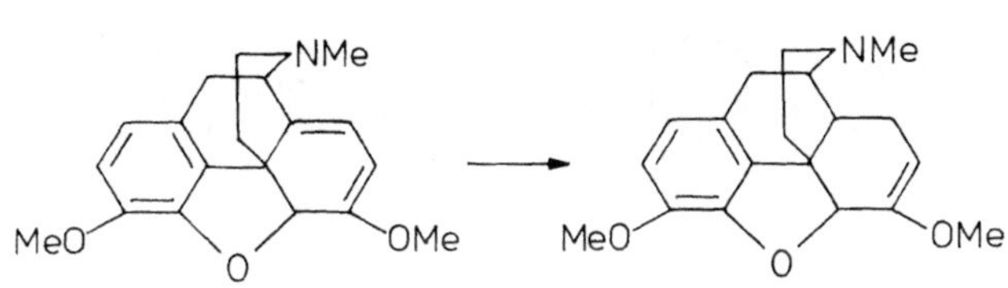

(49) (50)

4. Diene Systems in Steroids and Steroid Enones

The case of thebaine also exemplifies the selective hydrogenation of a 1,3-diene system. A further example [13] is provided by the hydrogenation of ergosterol (51) to the 5,6-dihydro derivative (52), i.e. the same result as in hydrogenation of ergosterol heterogeneously by means of hydrogen and platinum [25]. It is to be noted that the disubstituted 22,23-olefinic bond

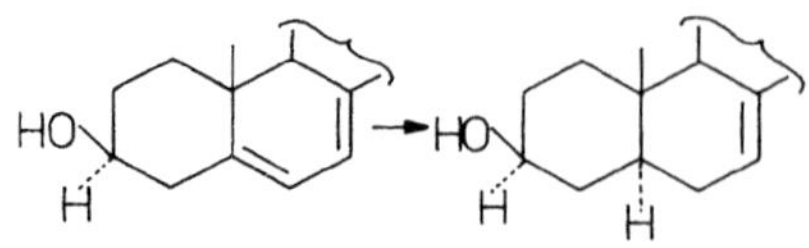

(51) (52)

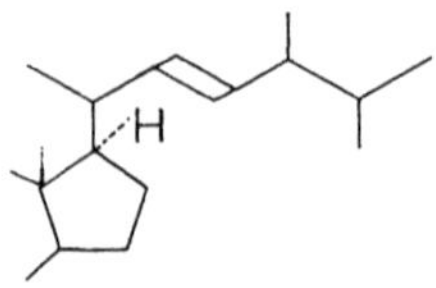

(53)

of the ergosterol side chain (as in 53) fails to hydrogenate at the same time as the diene system reacts.

Homogeneous catalysis has been very successfully applied to the selective hydrogenation of a range of steroid types. The steroid-1-en-3-one system as in (54) is readily hydrogenated [26]. Deuteriation (q.v.) of (54) established

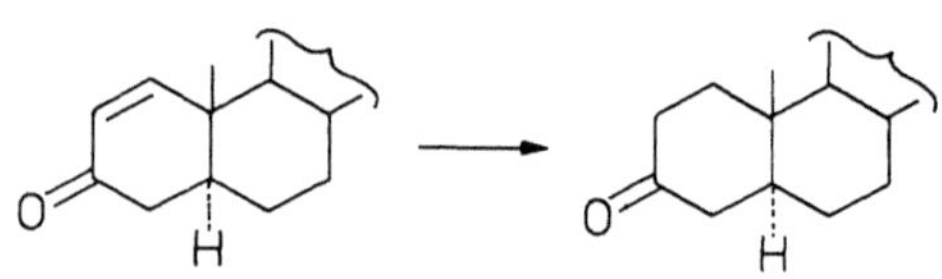

(54) (55)

that hydrogen addition occurs stereoselectively from the less hindered side of the steroid structure, i.e. in the sense of 1α,2α-addition.

In steroid 1,4-dien-3-ones the 4,5-olefinic centre proves, as expected, rather resistant to hydrogenation, and the first product from (56) arises by

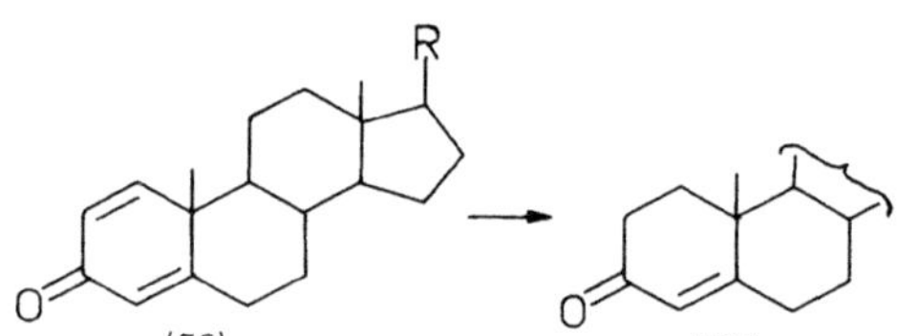

(56) (57)

preferential hydrogen addition to the 1,2-alkene bond. However, on prolonged hydrogenation (several days) the 4,5-unsaturated centre is also attacked leading, in the case of (56), R = 17β OH, to androstan-17βol-3-one [13] (58).

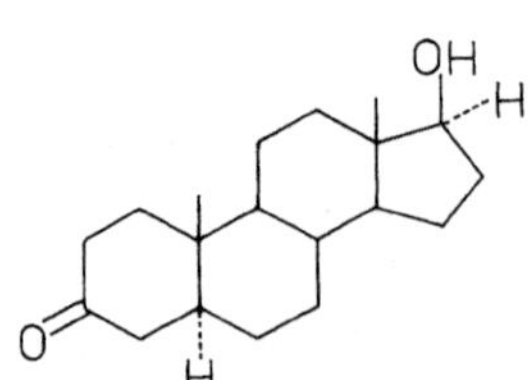

(58)

It is also observed that when hydrogen addition at the 4,5-alkene occurs the stereochemistry of hydrogen addition to the steroid-4-en-3-one system is a function of the nature of the substituent R in the 17-position (cf. 56). Data referring not to hydrogenation with $RhCl(PPh_3)_3$ as catalyst, but to a related system, will be discussed below. However, in the light of this information it is unsafe to generalise from the observed stereochemistry of hydrogen addition which leads to (58) as the principal isomer from (57).

The degree of discrimination between the 1,2- and 4,5-olefinic centres in a structure of type (56) is also, as expected, sensitive to steric factors in the remainder of the molecule. It was found, for example, that in (59), which is the 11α-acetoxy-20-ethylenedioxy analogue of (56) the 1,2- and 4,5-olefinic bonds are hydrogenated at rather similar rates using $RhCl(PPh_3)_3$ in benzene. The product contained the 1,2-dihydro- (60) and the tetrahydro-derivative (61) in about equal amounts [27], and clearly the 11α-acetoxy substituent introduces steric hindrance to co-ordination of the 1,2-alkene bond. However, 9,11-secoprogesterone was successfully prepared [27a] by RhCl $(PPh_3)_3$ catalysed hydrogenation of the corresponding 1,4-dien-3-one.

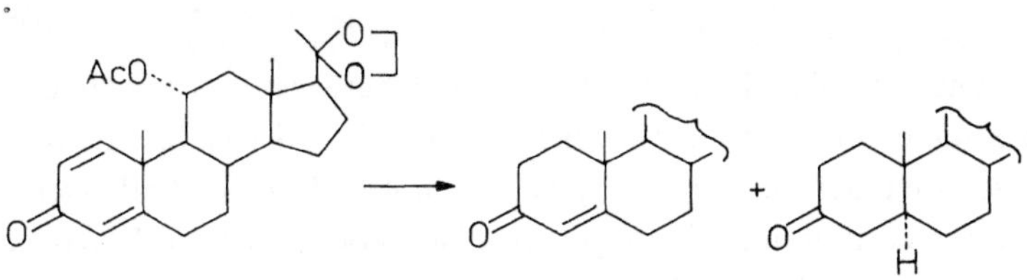

(59) (60) (61)

In the case of the steroid with the 4,6-diene-3-one structure the less substituted 6,7-olefinic centre is found to react preferentially, i.e. in the presence of $RhCl(PPh_3)_3$ in benzene-ethanol, (62) leads [26] to (63).

In general in steroids, olefinic centres at the 1,2-, 2,3-, 3,4- or 11,12-positions are hydrogenated without difficulty in presence of $RhCl(PPh_3)_3$ (Ref-

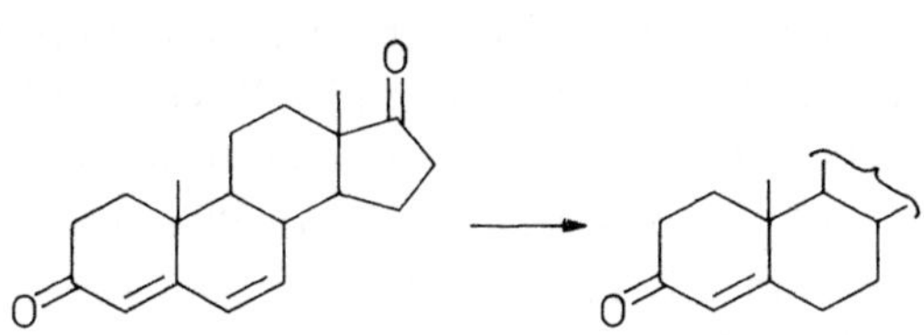

(62) (63)

erences 26, 16 and 28), but in the 5,6-position the double bond is quite resistant to hydrogenation [13]. This is illustrated by the hydrogenation [13] of pregna-5,16-dien-3βol-20-one (64) selectively at the 16,17-olefinic bond using $RhCl(PPh_3)_3$ in benzene.

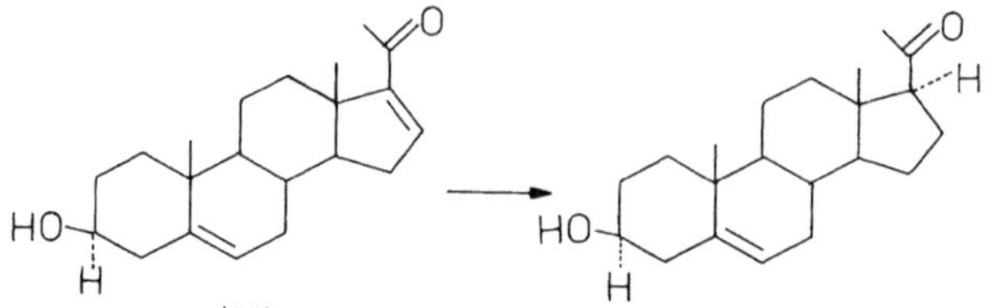

(64) (65)

The example (66)→(67), also effected using $RhCl(PPh_3)_3$ in benzene, illustrates an application [29] of homogeneous hydrogenation in the course

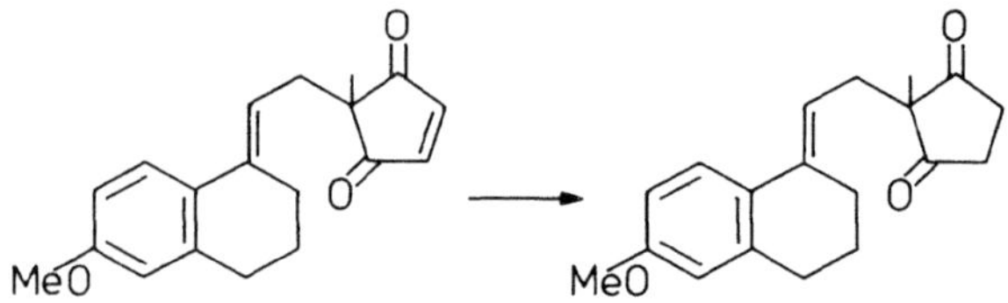

(66) (67)

of synthesis. The discrimination here is between an accessible disubstituted olefinic centre and a rather heavily substituted trisubstituted olefin.

A side reaction which may be encountered during hydrogenation of substances containing a keto group arises from a type of Lewis acid catalytic activity of the $RhCl(PPh_3)_3$ complex in promoting ketal formation with an alcoholic solvent [30]. This has been noted with 3-oxo-steroids. However,

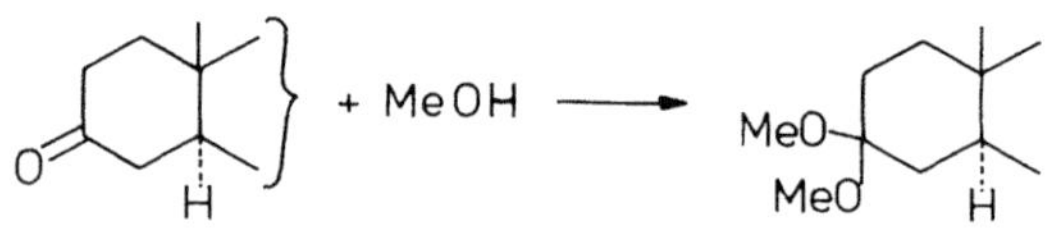

ketalisation is a relatively slow process, and it appears to be most effective with methanol solvent, and less effective with ethanol or *n*-propanol.

It was noted above that the ruthenium complex $RuHCl(PPh_3)_3$ is rather more selective than $RhCl(PPh_3)_3$ towards differently substituted alkenes. The hydrogenation of androsta-1,4-dien-3,17-dione (68) using $RuCl_2(PPh_3)_3$ in benzene has been examined [31a]. The effective catalyst is no doubt the hydrido complex, $RuHCl(PPh_3)_3$ formed *in situ* under hydrogen. The interest in the result in this case is that the proportions of the 1,2-dihydro product (69) and of the tetrahydro product (70) were found to depend on the hydrogen pressure. The amount of the dihydro product (69) was found to rise with the hydrogen pressure reaching 92.5% at 162 kg cm^{-2} H_2. However, the rate and selectivity of hydrogenation are influenced also by added bases [31b].

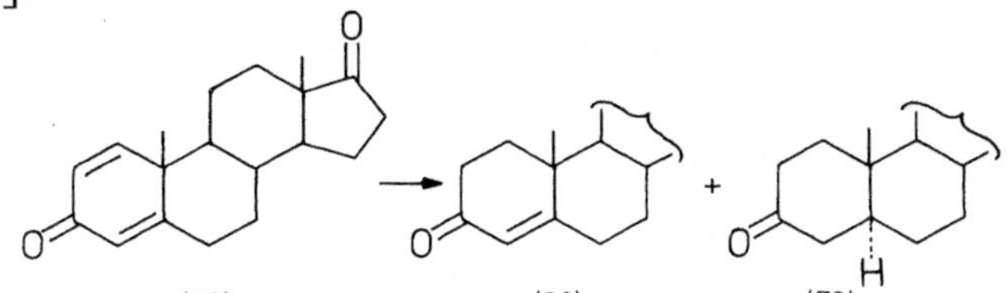

(68) (69) (70)

Although, as has been noted, steroid-4-en-3-ones have proved relatively resistant to hydrogenation by means of the $RhCl(PPh_3)_3$ catalyst, hydrogenation in the presence of the catalyst $RhCl_3py_3$—$NaBH_4$ in dimethylformamide occurs quite readily [32]. Also, employing this catalyst for hydrogenation of steroids of the type (71), it was possible to show that the proportions of the isomeric 5αH- and 5βH-hydrogenation products (72) and (73) are dependent on the nature of the remote 17-substituents, as shown. The proportions also differ from those found in heterogeneous hydrogenation of these steroids at rhodium on charcoal (proportions shown in parenthesis).

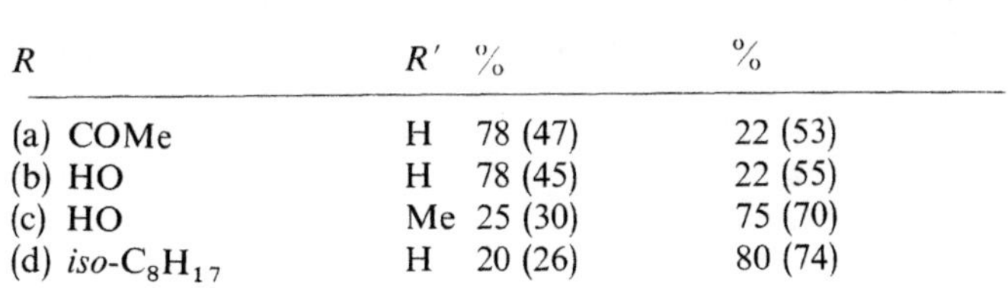

(71) (72) (73)

R	*R'*	%	%
(a) COMe	H	78 (47)	22 (53)
(b) HO	H	78 (45)	22 (55)
(c) HO	Me	25 (30)	75 (70)
(d) *iso*-C_8H_{17}	H	20 (26)	80 (74)

The examples (71) (b) and (c) both contain a polar hydroxyl 17-substituent i.e. polarity does not appear to be the decisive factor in determining the stereochemistry of hydrogen addition at the 4,5-bond. However, what does appear to be important is the total steric bulk of the substitution at 17 which transmits by steric compression through the structure a long range effect which at the 4,5-position influences the relative ease of co-ordination and hydrogenation from the lower or upper face of the molecule.

These interesting results also point to the use of homogeneous catalysis as a means of extending the range of conditions applicable to hydrogenation, and particularly in a direction which offers a means of improving the stereoselectivity of hydrogen addition.

The example noted above where the catalyst $RuCl_2(PPh_3)_3$ was employed with various pressures of hydrogen, however, also draws attention to the importance of controlling the kinetic circumstances, notably the catalyst and reactant concentrations and the hydrogen pressure. These factors may make a critical difference to reaction *via* alternative pathways differing little in activation energy, but leading to different stereoisomeric products. In the hydrogenation of cholestenone (71 (d)) using the $RhCl_3py_3$ catalyst the yield of the 5αH-product (72(d)) was indeed found [32] to fall somewhat with increasing steroid concentration.

5. Alkene Isomerisation

In considering the above examples of homogeneous hydrogenation catalysis no reference has been made to the phenomenon, common in heterogeneous catalysis, of isomerisation or translocation of the alkene bond. This complication which is discussed more fully in Chapter VII, is fortunately less frequently encountered in hydrogenation with most homogeneous catalytic complexes. Examples which illustrate this point are found in the hydrogenation [33] of the 1,4,5,8-tetrahydronaphthalene derivatives (74) or (75). Using $RhCl(PPh_3)_3$ as catalyst in benzene-ethanol, (74) was reported to yield (76),

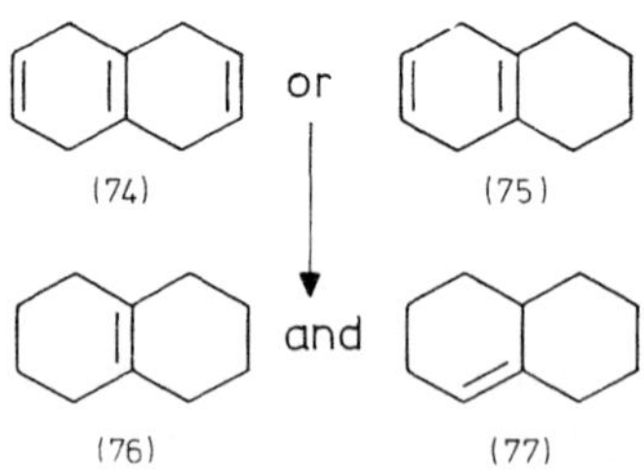

the expected product of hydrogenation of the less substituted olefinic bonds, accompanied by some 20% of an isomer (77). However, there was no report of isomerisation leading to aromatisation of the triene system of (74). Thus, in these examples, hydrogenation is fast relative to olefin bond migration. However, the rate difference may not be large since (78), a methoxycarbonyl substitution product of (74) was reported to give a little of the aromatic product (80) in addition to the main product (79) on hydrogenation at $RhCl(PPh_3)_3$ in benzene.

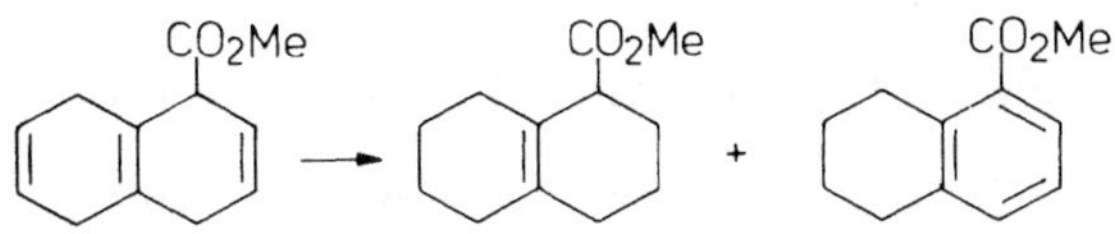

(78) (79) (80)

6. Allenes

The behaviour of a selection of molecules containing an allene grouping towards hydrogenation at $RhCl(PPh_3)_3$ in benzene has been examined [34]. Modest yields of derived *cis*-alkenes were reported, viz.:

$$CH_3(CH_2)_5CH{=}C{=}CH_2 \rightarrow CH_3(CH_2)_5CH{=}CHCH_3 \quad 66\%$$

$$CH_3(CH_2)_2CH{=}C{=}CH(CH_2)_2CH_3 \rightarrow$$

$$\rightarrow CH_3(CH_2)_3CH{=}CH(CH_2)_2CH_3 \quad 30\%$$

$$(CH_3CH_2)_2C{=}C{=}CH_2 \rightarrow (CH_3CH_2)_2C{=}CHCH_3 \quad 50\%$$

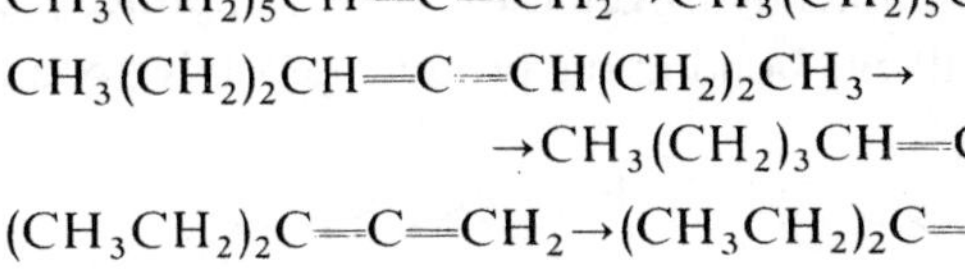

$n = 6$ 40%

$n = 10$ 65%

Hydrogenation appears to be slow, but no alkane or cycloalkane was reported, and the survival of the alkene products indicates the strong coordination of the allene. Allenes are, however, known to react readily with hydrido complexes. Thus allene with $RhH(CO)(PPh_3)_3$ or $IrH(CO)_2(PPh_3)_2$ forms [35] π-allyl derivatives (81) or (82).

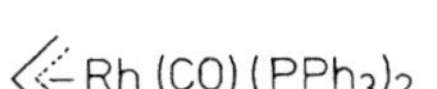

(81)

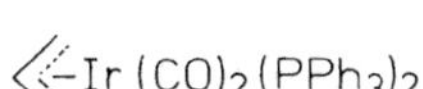

(82)

7. 1,3-Dienes

Catalysts which are specific for 1,3-dienes include pentacyanocobaltate (II) ion, $Co(CN)_5^{3-}$, and chromium tricarbonyl complexes such as $Cr(CO)_3 \cdot PhCO_2Me$ and related carbonyl derivatives.

Pentacyanocobaltate (II) ion has the disadvantage of being normally prepared in aqueous solution, and of showing an ageing process which results in loss of reactivity towards hydrogen. This ageing is at least partly due to reaction [36] with water:

$$Co(CN)_5^{3-} + H_2O \rightarrow HCo(CN)_5^{3-} + Co(CN)_5OH^{3-}$$

to give a species $Co(CN)_5(OH)^{3-}$ which does not react with hydrogen.

Alcoholic solutions of the lithium salt $Li_3CoH(CN)_5$ have been prepared [37] and shown to react with hydrogen. However, the salt, isolated as an ethanol solvate, proved rather sensitive to small amounts of water which cause formation of the dimeric $Li_6Co_2(CN)_{10}$.

Hydrogenation using pentacyanocobaltate (II) ion has been applied [38] to simple 1,3-dienes, e.g. buta-1,3-diene, and alkyl buta-1,3-dienes. However, extensive alkylation as in 2,5-dimethyl hexa-2,4-dien (83) introduces steric hindrance and inhibits hydrogenation. Also, as pointed out in Chapter I, the position of the alkene bond in the product depends upon the Co/CN^- ratio.

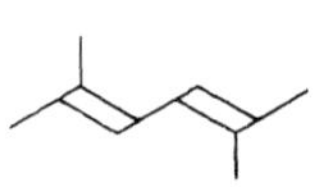

(83)

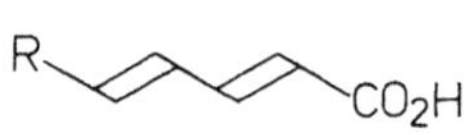

(84)

Cyclic dienes such as cyclopentadiene and cyclohexa-1,3-diene are reported to be hydrogenated by pentacyanocobaltate (II) to give the corresponding cycloalkene. However, larger cycloalkadienes or tri- and tetra-enes react rather slowly [39].

Various dienoic acids such as sorbic acid (84, R = Me) or muconic acid (84, $R = CO_2H$) have been also hydrogenated [40] using pentacyanocobaltate (II), as well as aryl conjugated monoenoic acids, [41], e.g. cinnamic acid (85), or atropic acid (86).

$$PhCH \cdot CHCO_2H \qquad\qquad PhC(=CH_2)CO_2H$$

(85) (86)

Maleic, fumaric and acrylic acids [42] seem somewhat resistant to hydrogenation, but reaction is effected at a somewhat higher temperature, e.g. 70°. N.m.r. examination of the adducts formed between crotonic or fumaric acid and $DCo(CN)_5^{3-}$ indicates a process of *cis*-addition [41], but similar reaction with citraconic (87), or mesaconic acid (88) appears not to be stereospecific [42].

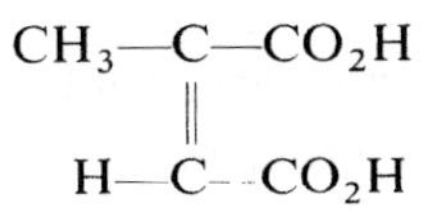
(87)

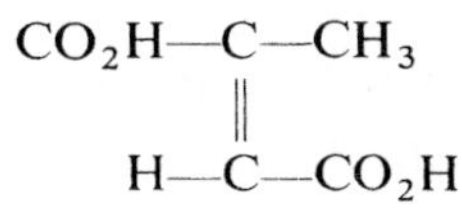
(88)

Various $\alpha\beta$-unsaturated aldehydes have been hydrogenated at the alkene centre using $Co(CN)_5^{3-}$, but, as is observed with simple alkene carboxylic acids, reduction does not occur readily [41].

Tricarbonylchromium-methyl benzoate used in a hydrocarbon solvent at 160° and 30 atm pressure of hydrogen has been applied [43] to the hydrogenation of a range of 1,3-dienes. The products appear to be largely, but not exclusively the result of 1,4-hydrogen addition and a number of examples are listed in Table II of Chapter I.

Using $Cr(CO)_3 \cdot PhCO_2Me$ as catalyst 1,4-dienes appear to undergo prior conjugation, and cyclohexa-1,4- and 1,3-dienes both yield cyclohexene [43]. Hence a 1,4-diene on deuteriation with this catalyst necessarily leads to a mixture of isomeric ^2H-derivatives [44].

A number of examples of the hydrogenation of conjugated dienes or dienones have been included above in the discussion of the hydrogenation of alkenes. These refer to the use of catalysts which unlike $Co(CN)_5^{3-}$ or $(PhCOOMe)Cr(CO)_3$ are not specific for hydrogenation of 1,3-dienes, and also to dienes in which the hydrogenation is largely directed by the pattern of substition. However, it is appropriate to refer to a detailed study [45] of the hydrogenation of buta-1,3-diene using triphenylphosphine iridium (I) carbonyl hydrides.

The catalyst involves both mono- and trihydrido species which are related by the equilibria:

$$IrH(CO)(PPh_3)_3 \rightleftarrows IrH(CO)(PPh_3)_2 + PPh_3$$
$$IrH(CO)(PPh_3)_2 + H_2 \rightleftarrows IrH_3(CO)(PPh_3)_2$$

With buta-1,3-diene both hydrides lead to a π-allyl complex, Ir(π − 1-methallyl) (CO) ($PPh_3)_2$ which can be isolated. This intermediate reacts with hydrogen to yield but-1-ene, *cis*- and *trans*-but-2-enes, and subsequently *n*-butane. Thus the pattern of reaction is very similar to that observed with $HCo(CN)_5^{3-}$.

8. Unsaturated Aldehydes

From many examples discussed in this chapter, it is clear that homogeneous hydrogenation of an alkene conjugated to a keto group is generally possible, the rate depending on the degree of substitution around the alkene bond. Although formally similar, unsaturated aldehydes may behave rather differently towards the usual catalysts for homogeneous hydrogenation. Saturation of the alkene centre may be realised, but the hydrogenation is usually accompanied by some degree of decarbonylation of the aldehyde so that yields are often less than quantitative. In addition the abstracted carbon monoxide becomes co-ordinated to the catalytic complexes, and RhCl $(PPh_3)_3$ for example gives rise to RhCl(CO) $(PPh_3)_2$ which does not activate hydrogen and the catalyst is in this way progressively deactivated.

Hydrogenation with $RhCl(PPh_3)_3$ is found [46] to be most effective if the unsaturated aldehyde is added in dilute solution to the catalyst presaturated with hydrogen. In this way it was possible to realise satisfactory hydrogenation of propenal, but-2-enal, and 2-methyl pent-2-enal to give the corresponding saturated aldehydes [46]. A small proportion of the alkanol may also at times be observed. For hydrogenation of cinnamaldehydes the decarbonylation reaction has been found [47] to be minimised when absolute ethanol is used as the solvent, but the yield of the 3-phenylpropanal was then of the order of only 60%. However, attempted hydrogenation of citral (89) failed [13], and the catalyst was recovered as RhCl(CO) $(PPh_3)_2$.

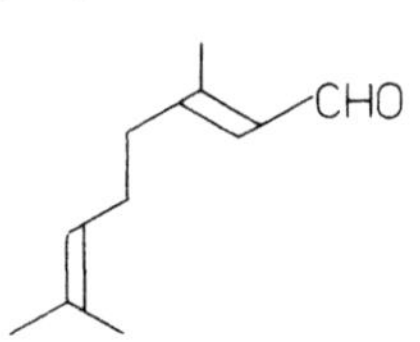

(89)

The material which is active in bringing about decarbonylation is the phosphine dissociated complex $RhCl(PPh_3)_2$ and the mechanism [48] of reaction may be outlined:

$$\mathrm{RhCl(PPh_3)_2} \xrightleftharpoons{\mathrm{RCHO}} \mathrm{RhCl(PPh_3)_2}(R\mathrm{CHO}) \rightleftarrows$$
$$\mathrm{(RCO)\,RhHCl(PPh_3)_2} \rightleftarrows \mathrm{RRh(CO)\,HCl(PPh_3)_2}$$
$$\rightleftarrows \mathrm{RhCl(CO)\,(PPh_3)_2} + \text{alkane, or alkene} + \mathrm{H_2}$$

This reaction sequence has indeed been made use of as a general method of converting aldehydes to alkanes [49], and the reaction is stereospecific [50].

It may be noted that the catalyst $RuCl_2(PPh_3)_3$ which is very active, particularly towards terminal alkenes, failed[46] to hydrogenate unsaturated aldehydes. A related complex, $[Ru_2Cl_3(PEt_2Ph)_6]Cl$ was found highly active in aldehyde decarbonylation [51].

References

1. I. Jardine and F. J. McQuillin: *J. Chem. Soc. C*, 458 (1966).
2. J. A. Osborn, F. H. Jardine, J. F. Young, and G. Wilkinson: *J. Chem. Soc. A*, 1711 (1966).
3. J. P. Candlin and A. R. Oldham: *Discuss. Faraday Soc.* **46**, 60 (1968).
4. G. C. Bond and R. A. Hillyard: *Discuss. Faraday Soc.* **46**, 20 (1968).
5. A. S. Hussey and Y. Takeuchi: *J. Amer. Chem. Soc.* **91**, 672 (1969); *J. Org. Chem.* **35**, 643 (1970).
6. F. H. Jardine, J. A. Osborn, and G. Wilkinson: *J. Chem. Soc. A*, 1974 (1967).
7. P. S. Hallman, B. R. McGarvey, and G. Wilkinson: *J. Chem. Soc. A*, 3143 (1968).
8. F. H. Jardine, J. A. Osborn, G. Wilkinson, and J. F. Young: *Chem. and Ind.*, 560 (1965).
9. C. O'Connor and G. Wilkinson: *J. Chem. Soc. A*, 2665 (1968).
10. W. Strohmeier and S. Hohmann: *Z. Naturforsch.* **25b**, 1309 (1970).
11. P. Abley, I. Jardine, and F. J. McQuillin: *J. Chem. Soc. C*, 840 (1971).
12. P. Abley and F. J. McQuillin: *J. Chem. Soc. Chem. Comm.*, 1503 (1969).
13. A. J. Birch and K. A. M. Walker: *J. Chem. Soc. C*, 1894 (1966).
14. E. Piers, R. W. Britton, and W. De Waal: *Canad. J. Chem.* **49**, 12 (1971).
15. A. Misono, Y. Uchida, M. Hidai, and T. Kuse: *J. Chem. Soc. Chem. Comm.*, 981 (1968); M. Brown and L. W. Piszkiewicz: *J. Org. Chem.* **32**, 2013 (1967).
16. J. F. Biellman and H. Liesenfelt: *Bull. Soc. chim. France*, 4029 (1966).
17. A. Tanaka, R. Tanaka, H. Uda, and A. Yoshimoshi: *J. Chem. Soc. Perk. I*, 1921 (1972).
18. H. C. Odom and A. R. Pinder: *J. Chem. Soc. Perk. I*, 2193 (1972).
19. J. E. McMurry and S. J. Isser: *J. Amer. Chem. Soc.* **94**, 7132 (1972).
20. J. J. Sims, V. K. Honwad, and L. H. Selman: *Tetrahedron Letters*, 87 (1969).
21. E. Piers and K. F. Cheng: *Canad. J. Chem.* **46**, 377 (1968).
22. A. J. Birch and K. A. M. Walker: *Australian J. Chem.* **24**, 513 (1971).

22a. M. Pardhasaradi and G. S. Sidhu: *Tetrahedron Letters*, 4201 (1972).

23. D. Gagnaire and P. Vottero: *Bull. Soc. chim. France*, 164 (1970).
24. A. B. Hornfeldt, J. S. Gronowitz, and G. Gronowitz: *Acta Chem. Scand.* **22**, 2725 (1968); A. J. Birch and K. A. M. Walker: *Tetrahedron Letters*, 1935 (1967).
25. I. M. Heilbron and W. A. Sexton: *J. Chem. Soc.*, 921 (1929); cf. L. F. and M. Fieser: *Steroids*, Rheinhold Pub. Corp., 1959, p. 112.
26. C. Djerassi and J. Gutzwiller: *J. Amer. Chem. Soc.* **88**, 4537 (1966).
27. P. Wieland and G. Anner: *Helv. Chim. Acta* **51**, 1698 (1968).
27a. N. S. Crossley and R. Dowell: *J. Chem. Soc. (C)*, 2496 (1971).
28. W. Voelter and C. Djerassi: *Chem. Ber.* **101**, 58 (1968).
29. R. D. Hoffsomer, D. Taub, and N. L. Wendler: *J. Org. Chem.* **32**, 3074 (1967).
30. W. Voelter and C. Djerassi: *Chem. Ber.* **101**, 1154 (1968).
31. S. Nishimura and K. Tsuneda: *Bull. Chem. Soc. Japan* **42**, 852 (1969).
31b. S. Nishimura, J. Ichino, A. Akimoto, and K. Tsuneda: *Bull. Chem. Soc. Japan* **46**, 279 (1973).
32. I. Jardine and F. J. McQuillin: *J. Chem. Soc. Chem. Comm.*, 503 (1969); *J. Chem. Soc. (C)*, 840 (1971).
33. J. J. Sims, V. K. Honwad, and L. H. Selman: *Tetrahedron Letters*, 87 (1969); E. Piers and K. F. Cheng: *Canad. J. Chem.* **46**, 377 (1968).
34. M. M. Bhagwat and D. Devaprabhakara: *Tetrahedron Letters*, 1391 (1972).
35. C. K. Brown, W. Mowat, G. Yagupsky, and G. Wilkinson: *J. Chem. Soc. (A)*, 850 (1971).
36. N. K. King and M. E. Wingfield: *J. Amer. Chem. Soc.* **83**, 2517 (1961).
37. G. Pregaglia, D. Morelli, F. Conti, G. Gregorio, and R. Ugo: *Discuss. Faraday Soc.* **40**, 110 (1968).
38. J. Kwiatek and J. K. Seyler: *Adv. in Chem. Series* **70**, 207 (1968).
39. W. Strohmeier and N. Iglauer: *Z. physik. Chem.* **61**, 29 (1968).
40. L. Simandi, F. Nagy, and E. Budo: *Acta Chim. Acad. Sci. Hung.* **58**, 39 (1968); L. Simandi and E. Budo: *ibid.* **64**, 125 (1970).
41. J. Kwiatek, I. L. Mador, and J. K. Seyler: *Adv. in Chem. Series* **37**, 201 (1963).
42. L. M. Jackman, J. A. Hamilton, and J. M. Lawlor: *J. Amer. Chem. Soc.* **90**, 1914 (1968); M. Murakami, R. Kawai, and K. Suzuki: *J. Chem. Soc. Japan* **84**, 1669 (1963).
43. E. N. Frankel and R. A. Butterfield: *J. Org. Chem.* **34**, 3930 (1969).
44. E. N. Frankel, E. Selke, and C. A. Glass: *ibid.* **34**, 3936 (1969).
45. H. G. Burnett, R. G. Morrison, and C. J. Strugnell: *J. Chem. Soc. Dalton Trans.*, 1663 (1974).
46. F. H. Jardine and G. Wilkinson: *J. Chem. Soc. (C)*, 270 (1967).
47. R. E. Harmon, J. L. Parsons, D. W. Cooke, S. K. Gupta, and J. Schoolenberg: *J. Org. Chem.* **34**, 3684 (1969).
48. M. C. Baird, C. J. Nyman, and G. Wilkinson: *J. Chem. Soc. (A)*, 348 (1968).
49. J. Tsuji and K. Ohno: *J. Amer. Chem. Soc.* **88**, 3452 (1966); **90**, 99 (1968).
50. H. M. Walborsky and L. E. Allen: *Tetrahedron Letters*, 823 (1970).
51. R. H. Prince and K. A. Raspin: *J. Chem. Soc. (A)*, 612 (1969).

CHAPTER IV

HYDROGENOLYSIS OF CYCLOPROPANES

The cyclopropane ring is generally rather stable to hydrogenolysis and hydrogenolytic fission of cyclopropanes by heterogeneous catalysis is commonly found to be quite slow. Cyclopropanes are equally stable towards homogeneously catalysed hydrogenolysis, and in the studies so far reported cyclopropane hydrogenolysis by homogeneous catalysis is observed only in cases where the cyclopropyl residue is in conjugation with an alkene bond, and then to only a limited extent. Using $RhCl(PPh_3)_3$ in benzene, hydrogenolysis has been observed in the examples (1)–(6), the extent of hydrogenolysis and the nature of the hydrogenolysis products being as indicated [1].

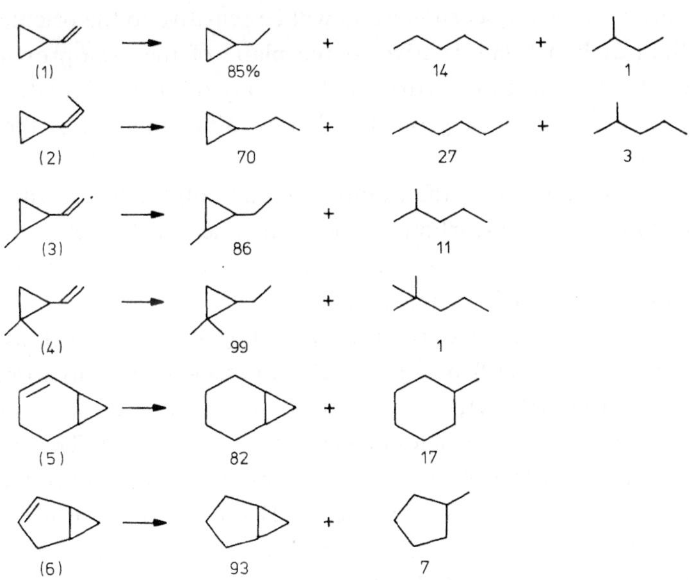

The intervention of hydrogenolysis is rationalised in terms of an addition-elimination process, similar to that invoked in alkene isomerisation (Chapter

VII), with the difference that the eliminated fragment arises by rupture of a cyclopropane C—C bond rather than a C—H bond, viz.:

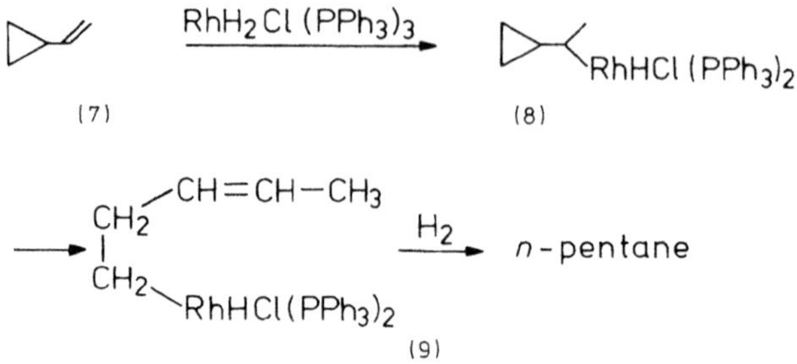

The parallel with olefin isomerisation is clearly evident.

It follows therefore that the extent of cyclopropane hydrogenolysis will be dependent on the rate of hydrogenolysis of the intermediate rhodium alkyl (8) relative to the rate of rearrangement to (9). Rearrangement will, however, also involve a geometrical factor since the step involving insertion into the cyclopropane carbon-carbon bond will be sensitive to the orientation of the rhodium-carbon bond relative to the plane of the cyclopropane ring. In different cases therefore hydrogenolysis may occur to a greater or smaller extent depending on the geometrical constraints imposed by the molecular structure.

It is of interest that in the examples (3) and (4) the less substituted of the cyclopropane carbon-carbon bonds is the one which is cleaved. A similar preference for the less substituted bond is found in the formation of $PtCl_2$ insertion products of cyclopropanes [2].

In more complex cyclopropane-containing structures such as the 1α,5α-cyclocholest-2-ene (10) or 3α,5α-cyclocholest-6-ene (12) hydrogenation by means of $RhCl(PPh_3)_3$ in benzene gave [3] the corresponding cyclocholestane (11) or (13) in yields of 80 and 90%, respectively. The authors made no mention of hydrogenolysis products, but the recorded products (11) and (13) were purified chromatographically. However, the recorded yields [4] of the cubebanones (15) and (17), derived from (14) and (16) respectively by hydrogenation using $RhCl(PPh_3)_3$ in benzene, are such as to indicate no cyclopropane hydrogenolysis in these cases. In (14) and (16) the carbonyl group but not the alkene centre is in conjugation with the cyclopropane ring.

(10) (11)

(12) (13)

(14) (15)

(16) (17)

However, as has already been noted the $RhH_2Cl(PPh_3)_2$ hydride catalyst does not interact with the carbonyl group. Thus (14) and (15) would not be expected to show cyclopropane hydrogenolysis.

References

1. C. H. Heathcock and S. R. Poulter: *Tetrahedron Letters*, 2755 (1969).
2. F. J. McQuillin and K. G. Powell: *J. Chem. Soc. Dalton Trans.*, 2123 (1972).
3. S. B. Laing and P. J. Sykes: *J. Chem. Soc. (C)*, 421 (1968).
4. A. Tanaka, R. Tanaka, H. Uda, and A. Yoshimoshi: *J. Chem. Soc. Perk I*, 1721 (1972).

CHAPTER V

HYDROGENATION OF AROMATICS AND HETEROAROMATICS

The common use of benzene as solvent in homogeneous hydrogenation with complexes such as $RhCl(PPh_3)_3$, or $RuHCl(PPh_3)_3$ indicates the general resistance of the aromatic ring to hydrogen under these conditions. Also the survey in Chapter III of alkenes which have been hydrogenated by homogeneous catalysis included a number of examples of molecules containing a benzenoid or thiophene ring.

Various catalyst systems capable of hydrogenating aromatics or heteroaromatics have, however, been reported. Hydrogenation of N-heteroaromatics clearly presents the easier problem, and the catalyst obtained by treating $RhCl_3py_3$ in dimethylformamide with an equivalent of sodium borohydride was found [1] to catalyse the hydrogenation of pyridine and of quinoline in the heteroaromatic ring, viz.:

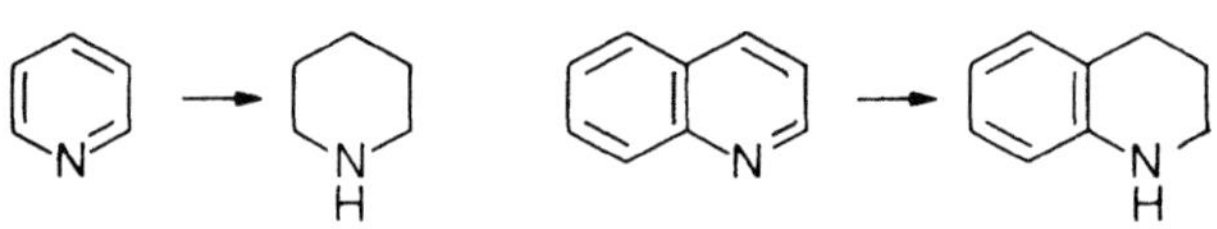

(1) (2) (3) (4)

The benzenoid ring in (3) or (4) was not hydrogenated, and for hydrogenation of the heteroaromatic ring hydrogen uptake required some hours, albeit at ambient temperature, and using only 1 atm pressure of hydrogen.

Since pyridine acts as a ligand in the active complex in this catalyts, i.e. $[RhCl_2py_2(dmf)BH_4]$, it is possible to envisage a process of hydrogenation being initiated by hydrogen transfer:

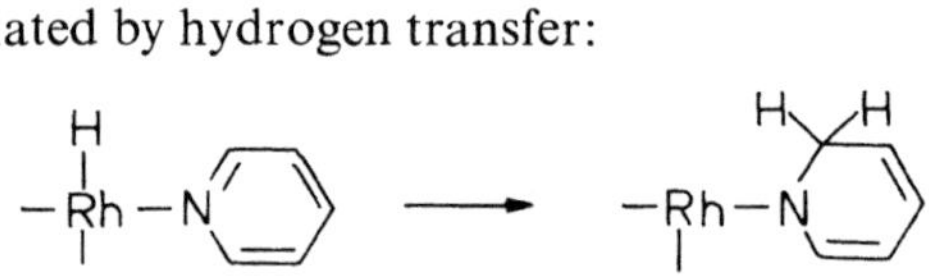

with subsequent hydrogen addition to this hydropyridine complex. An analogous process for a non-heteroaromatic is clearly less easily possible.

Nevertheless a group of complexes formulated as in (5) has been reported [2] by Russian workers to catalyse hydrogenation of aromatics in dimethylformamide solution at 20° and 1 atm H_2.

$H[Rh_2ClL_2]$ L=$PhCH_2COO^-$, 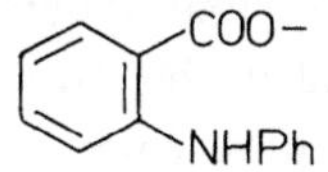

, or 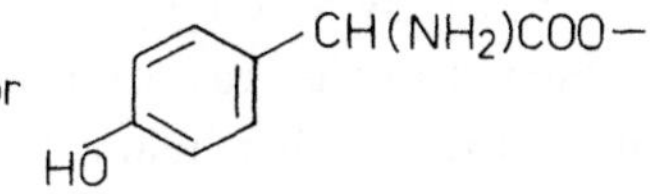

(5)

Examples of substances hydrogenated by catalysts of type (5) include the following:

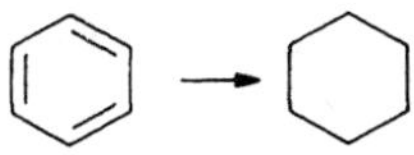

(6)

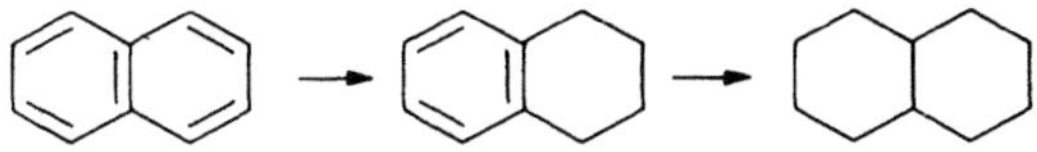

(7)

Unfortunately details of this interesting catalytic system are scarce. Alkenes are also hydrogenated using this catalyst, but, curiously, the carbonyl group is found not to be reduced [3], e.g. $Me_2C{=}CHCOCH_3 \rightarrow Me_2CHCH_2COCH_3$.

For industrial application very extensive studies have also been made of various Zeigler type combinations which have proved effective for hydrogenation of aromatics as well as simple alkenes and dienes. These systems although highly active consist only partly of soluble material, and the components are difficult to characterise.

Catalysts of this type have been derived from the action of e.g. triethyl aluminium on transition metal derivatives such as $M(acac)_2$, (M = Ni, Mn, Cu), $M(acac)_3$, (M = Fe, Co). $M(OCOCH_2CH(C_2H_5)CH_2C_2H_5)_2$ (M = Ni, Co), Ni (dimethyl glyoxime)$_2$, cp_2TiCl_2, usually in a hydrocarbon solvent such as pentane, heptane or cyclohexane.

The following aromatics are reported [4] to have been hydrogenated by

(8)

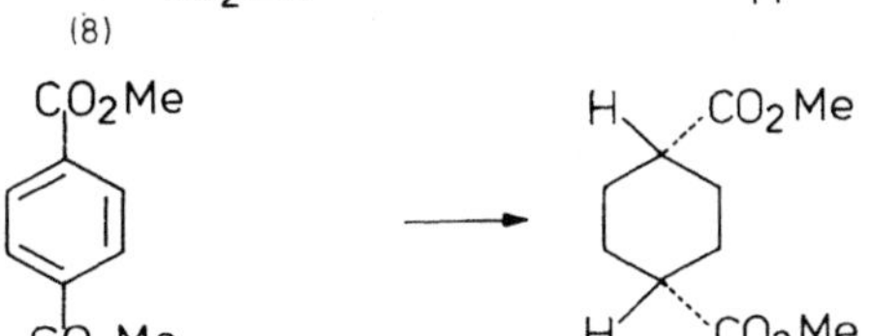

(9)

Zeigler type catalysts, benzene, *o*-, *m*-, and *p*-xylenes, naphthalene, pyridine, phenol to cyclohexanol, nitrobenzene to aniline and to cyclohexyl- and dicyclohexyl amine, and the phthalate esters (8) and (9).

The catalytic activity of these systems is dependent on the groups originally attached to the transition metal in the parent metal complex, and may be modified by other ligand molecules which are added, e.g. amines, phosphines, phosphites.

The mode of action of the group of catalysts is not quite clear, but presumably the aluminium ethyl generates transition metal ethyl derivatives. These are known to be sensitive to decomposition by β-elimination to form a hydride:

$$—M-CH_2CH_3 \rightarrow —MH + C_2H_4$$

However, it is evident that hydrogenation must also involve hydrogenolysis of the metal alkyl:

$$—M-\text{alkyl} + H_2 \rightarrow —MH + \text{alkane}$$

or metal cyclohexadienyl when an aromatic is hydrogenated.

A better defined catalyst system capable of effecting hydrogenation of benzene and other aromatics ate 25° and 1 atm H_2 has also recently been reported [5]. This makes use of a π-allyl trimethylphosphite cobalt, η^3-C_3H_5-$Co[P(OMe)_3]_3$. Examples of aromatics which have been hydro-

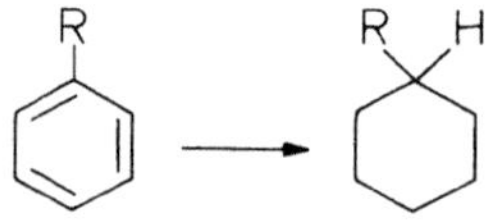

(10), R = H or alkyl

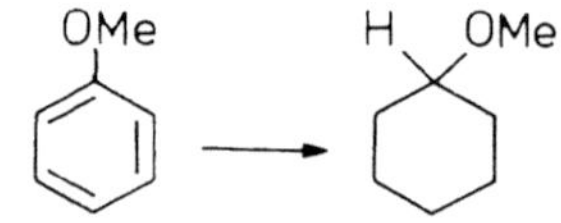

(11)

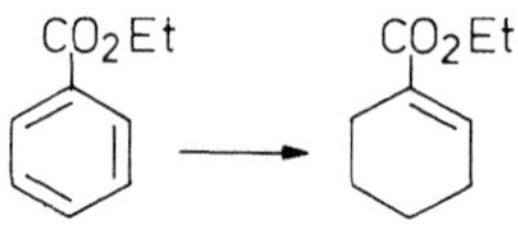

(12)

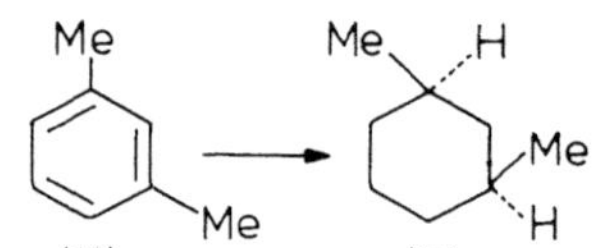

(13) (14)

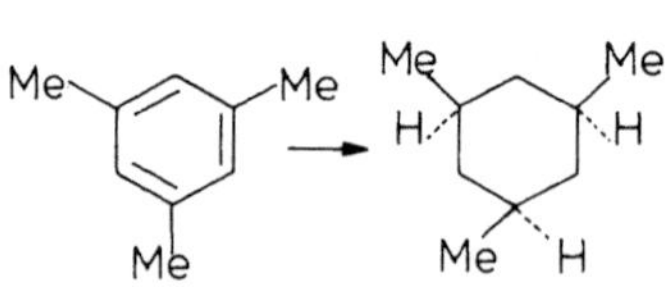

(15) (16)

(17)

genated include (10)–(13), (15) and (17). These results draw attention to the versatility of this catalyst and the reported stereospecific hydrogenation of (13) to the *cis*-di-methyl cyclohexane and of (15) to trimethylcyclohexane is noteworthy.

References

1. I. Jardine and F. J. McQuillin: *J. Chem. Soc. Chem. Comm.*, 626 (1970); P. Abley, I. Jardine, and F. J. McQuillin: *J. Chem. Soc. (C)*, 840 (1971).
2. O. N. Efimov, O. N. Eremenko, A. G. Ovcharenko, M. L. Khidekel, and P. S. Chekrii: *Bull. Acad. Sci. U.S.S.R. Chem. S.*, 778 (1969).
3. N. V. Borunova, L. Kh. Friedlin, M. L. Khidekel, S. S. Danielova, V. A. Avilov, and P. S. Chekrii: *Bull. Acad. Sci. U.S.S.R. Chem. S.*, 432 (1968).
4. Cf. B. R. James: *Homogeneous Hydrogenation*, Wiley-Interscience, New York, 1973, pp. 363–383.
5. E. L. Muetterties and F. J. Hirsekorn: *J. Amer. Chem. Soc.* **96**, 4063 (1974); F. J. Hirsekorn, M. C. Rakowski, and E. L. Muetterties: *J. Amer. Chem. Soc.* **97**, 237 (1975).

CHAPTER VI

HYDROGENATION OF MOLECULES CONTAINING CARBONYL, NITRO, HALOGEN, —N═N- AND —N═CH-GROUPS

1. The Carbonyl Group

It is commonly observed in heterogeneous hydrogenation that whereas alkenes and similar centres react readily, hydrogenation of ketones or aldehydes is slow and not infrequently subject to catalysis by use of an acidic medium. The difference corresponds with the fact that alkene or alkyne π-complexes are numerous and well characterised, whilst similar complexes involving the $>C{=}O$ group are few; in the aldehyde complexes [1] $[M(OCHR)_6](FeCl_4)_2$, where $M = Mg$, Mn, Fe, Co, Ni and $R = Me$, Et, Ph, the carbonyl group is acting as a Lewis base. Carbonyl complexes analogous to those formed by alkenes have, however, been obtained from hexafluoroacetone in which the acceptor capacity of the $>CO$ group is very much enhanced, and ν_{CO} is moved to 1810 cm^{-1}. Examples [2] include (1), $M = Pt$, Pd,

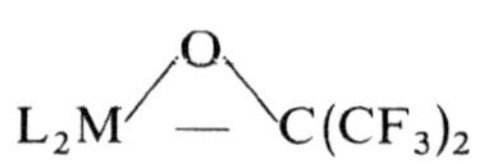

(1)

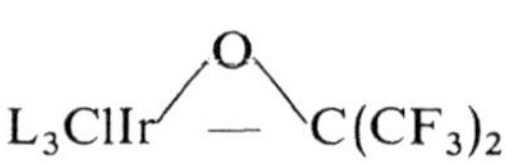

(2)

$L = Ph_3P$, $M = Pt$, $L = Ph_2MeP$, (2), $L_3 = (Ph_3P)_2$, CO; $(Ph_2PMe)_2$, CO, (3) and (4), L = various phosphines, phosphites, dipyridyl.

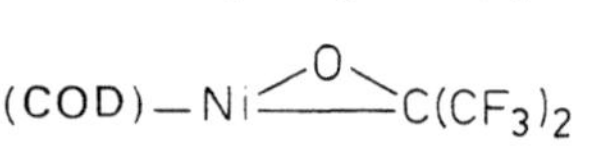

(3)

$L_2Ni[CO(CF_3)_2]$

(4)

From the range of carbonyl substituted molecules discussed in Chapter III, it will be evident that the carbonyl group is essentially inert to hydrogenation by catalytic complexes such as $RhCl(PPh_3)_3$, $RuHCl(PPh_3)_3$ which discriminate strongly in favour of co-ordination to alkene or alkyne centres. Also in the $RhCl_3py_3$-$NaBH_4$ catalyst system the BH_4 residue fails to exhibit its usual reactivity towards a carbonyl centre.

It appears therefore that for successful catalytic hydrogenation of carbonyl compounds the catalyst should exhibit a strongly nucleophilic hydride character, and enhanced acceptor capacity so as to assist co-ordination of the carbonyl group.

These factors appear to underline the presently known examples of carbonyl group hydrogenation by homogeneous catalysts.

It is a characteristic of pentacyanohydridocobaltate (III) ion to transfer hydrogen to the β-position of a conjugated enone system, i.e. to behave as a source of nucleophilic hydrogen. Although this catalyst hydrogenates the olefinic centre preferentially [3], as in the examples (5)–(6), sufficiently electrophilic ketones have been reduced at the carbonyl group [4], e.g. benzil (9), diacetyl (10) or *p*-benzoquinone (11).

$$RCOC(Me){:}CH_2 \rightarrow RCOCHMe_2$$

(5) R = Me, Ph (7)

$$PhCH{:}CHCOMe \rightarrow PhCH_2CH_2COMe$$

(6) (8)

(9) $PhCOCOPh \rightarrow PhCH(OH)\,COPh$

(10) $MeCOCOMe \rightarrow MeCH(OH)\,COMe$

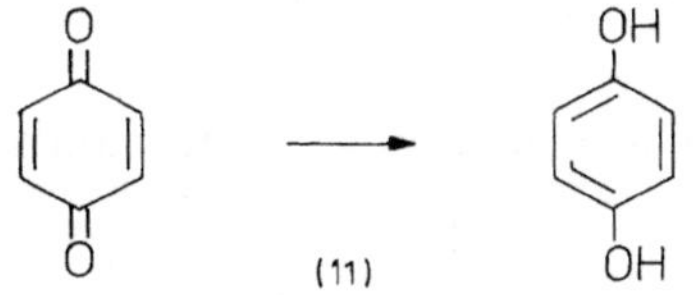

(11)

Rather similar reactivity is exhibited by the $RhCl_3py_3$—$NaBH_4$—H_2 catalyst in dimethylformamide, and slow hydrogenation of acetophenone (12), benzophenone (13), and benzoin (14) over a period of 1–3 days has been observed [5] with this catalyst.

$PhCOMe$,	Ph_2CO,	$PhCOCH(OH)Ph$
(12)	(13)	(14)

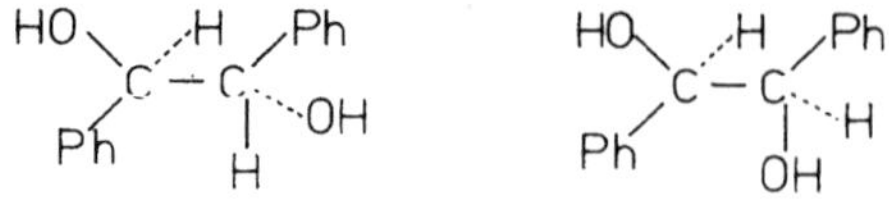

(15a) (15b)

Benzoin gave the usual mixture of meso- (85%) (15a) and (±)-hydrobenzoin (15%) (15b). However, cyclohexanone could not be hydrogenated with this catalyst, and it is evident that the borohydride used in preparing the catalyst makes no contribution to these results.

Both aliphatic and alicyclic ketones have been hydrogenated [6] using a cationic complex $[RhH_2(PR_3)_2 \cdot S_2]^+$ when S represents a solvent molecule such as ethanol, dioxan, dimethylformamide. This activity is, however, somewhat dependent on the phosphine ligand, PR_3. Whereas a triphenylphosphine complex will hydrogenate alkenes, dienes or acetylenes, hydrogenation of the carbonyl group requires the use of a more basic alkyl substituted phosphine. Using $[RhH_2(PPhMe_2)_2S_2]^+$ as the ClO_4^- or PF_6^- salt it was possible to observe rapid hydrogenation of acetone at 25° and 1 atm H_2, and of butan-2-one, cyclohexanone and acetophenone. 4-*t*-Butylcyclohexanone (16) gave the *trans*- (17) and *cis*-4-*t*-butylcyclohexanols (18) in a ratio of 86:14, i.e. hydrogen transfer occurs preferentially in the axial sense.

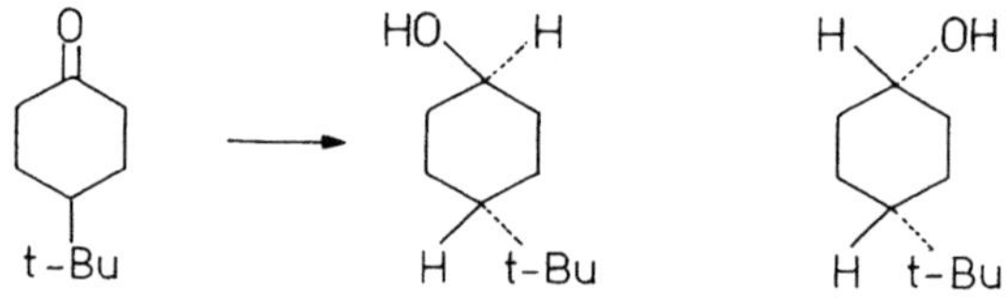

(16) (17) (18)

The following mechanism for hydrogenation with this catalyst has been outlined:

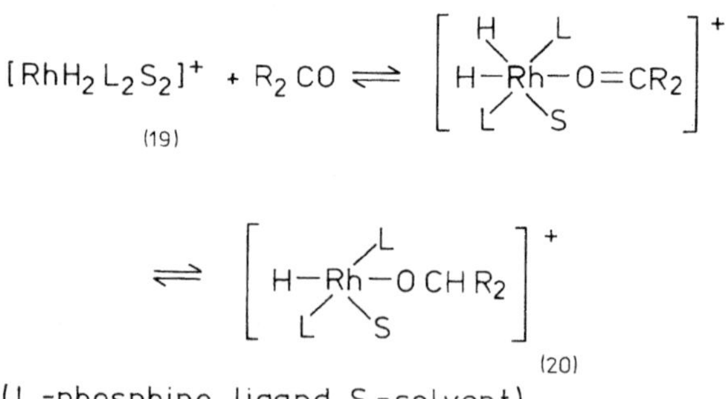

(19) (20)

(L = phosphine ligand, S = solvent)

and, since a small amount (1%) of water in the medium is markedly catalytic, the alkoxy complex (20) is thought [6] to suffer hydrolysis *via* e.g. (21).

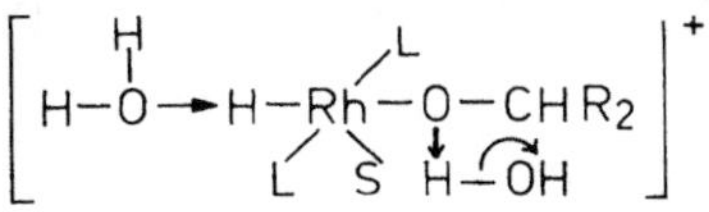
(21)

Iridium complexes have also been particularly explored in connexion with the hydrogenation of the carbonyl group. It was found that the trihydrido complex, $IrH_3(PPh_3)_3$ in acetic acid at 50° catalyses the hydrogenation of *n*-butyraldehyde at 1 atm H_2 and fairly rapidly [7]. It is likely that the effective agent is one or other of the hydrido-acetates such as $IrH_2(OAc)(PPh_3)_3$, but it is of interest that the catalyst is not deactivated due to decarbonylation of the aldehyde, as is found with $RhCl(PPh_3)_3$ (cf. Chapter III).

2. Carbonyl Reduction *via* Addition of a Silane

An indirect procedure for hydrogenation of the carbonyl group makes use of addition of a hydrosilane to the carbonyl group, catalysed by the $RhCl(PPh_3)_3$ complex [8]. It has been possible to characterise [8] an adduct $Rh(H)(Cl)(SiEt_3)(PPh_3)_2$ from reaction of $RhCl(PPh_3)_3$ with Et_3SiH, and this type of silyl rhodium hydride is presumably the effective reagent. The overall reaction leads to the silyloxy derivative which may be hydrolysed by means of aqueous methanolic potassium hydroxide. Reaction is generally rapid at the ordinary temperature, although aryl

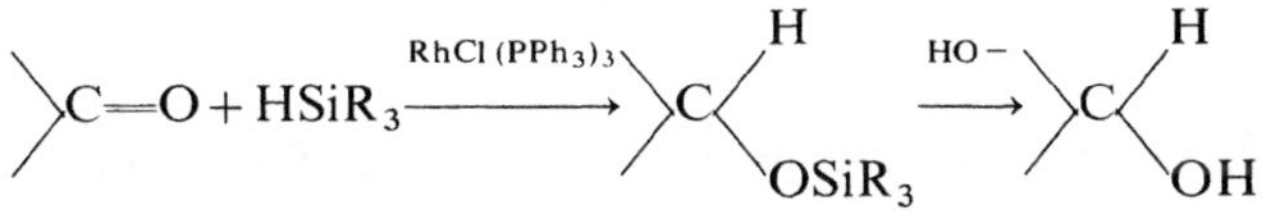

ketones react more slowly than alkyl ketones. Di- or tri-hydrosilanes react more rapidly than monohydrido silanes.

Examples of ketones and aldehydes which have been shown [8] to react in this way include PhCHO, PhCOMe, PhCOPh, Me_2CO, $CH_3CH_2CH_2CHO$, and cyclohexanone, and a range of silanes such as $PhMe_2SiH$, Ph_3SiH, Et_2SiH_2, $PhMeSiH_2$, Ph_2SiH_2, $PhSiH_3$ has been employed.

The stereochemistry of hydrosilylation has been examined [9] for camphor (22) and menthone (23). In general the predominant alcohol product (24) or (26), is that formed by hydrogen transfer from the least hindered direction,

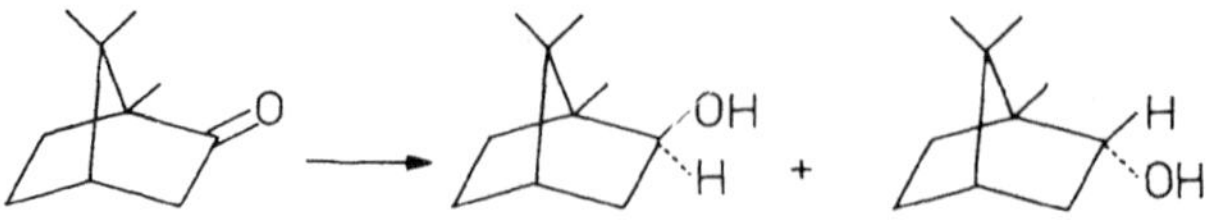

(22) (24) (25)

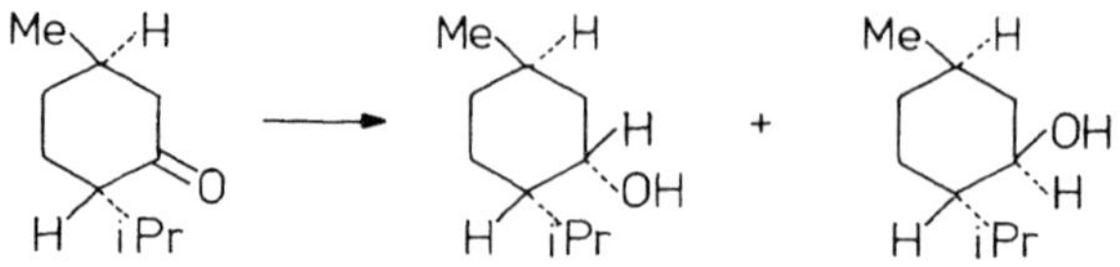

(23) (26) (27)

if a not too bulky silane is employed, e.g. $PhSiH_3$, Et_2SiH_2, Ph_2SiH_2, but increasing amounts of (25) or (27) result when a more bulky silane such as Et_3SiH is used.

In the case of α,β-unsaturated aldehydes or ketones the catalysed addition involves the olefinic bond, leading to a silylenol ether (28) which may be hydrolysed to the saturated aldehyde or ketone [8]:

$$\begin{matrix} R^1 \\ \\ R^2 \end{matrix}\!\!>\!C{=}CHCOR^3 \rightarrow \begin{matrix} R^1 \\ \\ R^2 \end{matrix}\!\!>\!CH{-}CH{=}C\!\!<\!\begin{matrix} R^3 \\ \\ OSi{\equiv} \end{matrix} \qquad (28)$$

$$\rightarrow R^1R^2CHCH_2COR^3$$

Examples of $\alpha\beta$-unsaturated carbonyl compounds which have been examined [8] include cinnamaldehyde, mesityl oxide, crotonaldehyde, and benzylidene acetone.

3. Carbonyl Reduction *via* Hydride Transfer from a Solvent

A catalytic procedure which gives slow, but highly stereoselective reduction of the carbonyl group employs a solution of sodium hexachloroiridate (IV) or chloroiridic acid in somewhat aqueous isopropanol in the presence of a phosphite such as triethylphosphite [10]. The effective catalyst is presumably a hydrido iridium phosphite complex formed in solution. This catalyst system is especially useful for the reduction of cyclohexanones to give the axial cyclohexanols with a high degree of stereoselectivity, i.e. by hydrogen addition from the less hindered equatorial direction. This steric preference is ascribed to the hydride transferring complex being rather bulky.

This catalyst system has been applied particularly to steroid ketones [11] where it is found to be highly specific for reduction of the 3-oxo group, oxo-groups at positions 6, 7, 11, 12 and 17 being apparently too sterically hindered for reduction. The following results [11] indicate also the high level of stereoselectivity of this catalyst in effecting reduction to the axial steroid alcohol, i.e. the 3α-hydroxy steroid, in the 5αH-, (29), and the 3β-hydroxy steroid in the 5βH-series (30). The figures for the percentages of the epimeric sterols indicate incomplete reduction in some cases, but there is very generally a considerable preponderance of the axially oriented epimer.

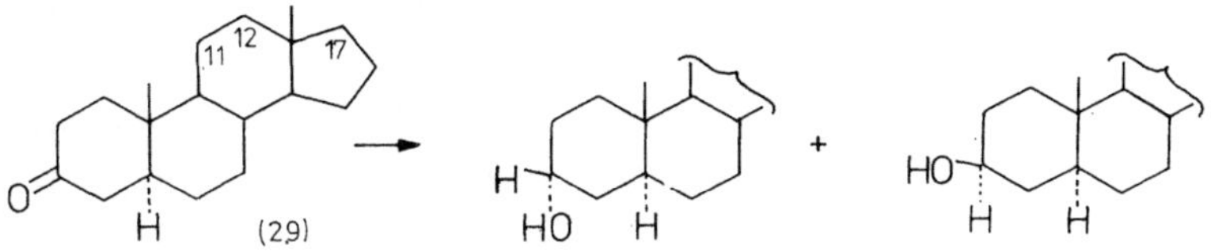

	90%	10%
17β OH	95	2
17 CO	94	2
11,17 diketo	94	5
17β COMe	78	7
17β CH(OH) Me	59	4
11 CO, 17β COMe	63	9
12 CO, 17β COMe	86	9
17 $CHMe(CH_2)_3CHMe$	68	29

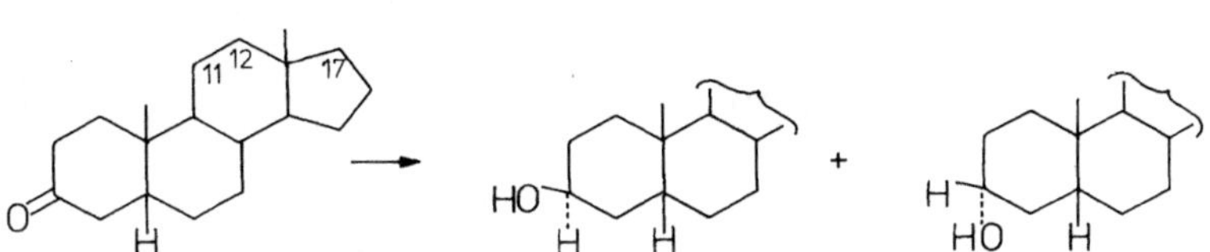

17 CO	98%	2%
17β OH	98	2
17 COMe	72	1
11 CO, 17 COMe	72	1
12 CO, 17 COMe	88	4

The process of reduction, however, is slow and requires prolonged heating under reflux, e.g. for up to 200 h. Under these conditions there may be some isomerisation of a 17β-COMe group leading to a 17α, 17β-COMe, mixture of isomers. A pregnan-20α-ol type steroid appears to be unaffected, but a pregnan-20β-ol steroid is sensitive to isomerisation to other products in presence of this catalyst. However, since it is possible to protect the 17-COMe moiety as the ethylene ketal, or the cortisone side chain as the *bis*-methylene dioxy derivative keto steroids of this type may be reduced without unwanted side reactions.

Examples of reduction of simpler systems which emphasise the stereoselectivity of the reagent include [12] the cyclohexanones (31)–(34).

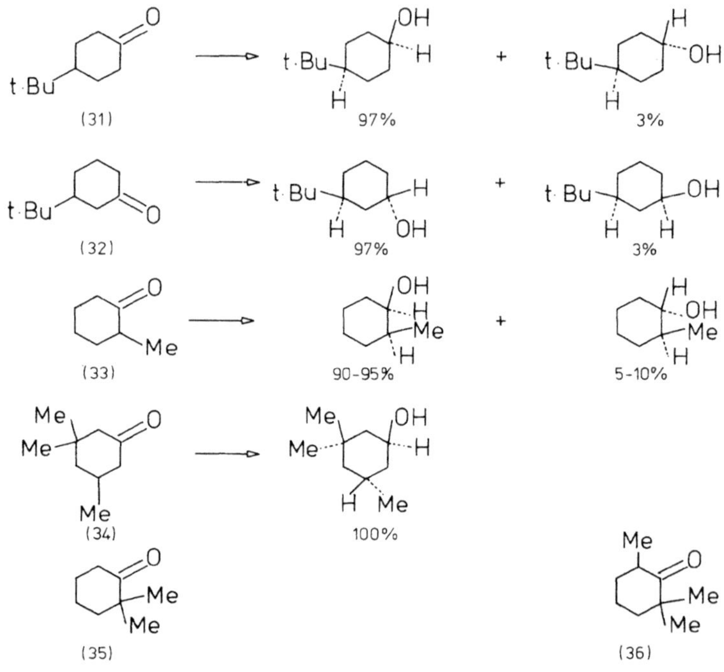

However, it may be noted that dialkyl substitution α to the carbonyl group, as in (35) and (36) effectively inhibits reduction.

An alternative catalytic system [13], also based on the use of chloroiridate (IV) salts or the acid M_2IrCl_6, M = H, K, NH_4, may be employed to catalyse hydrogen transfer between a cycloalkanone and an acyclic alkanol:

$$\text{cyclo—C=O} + \text{acyclic—}\overset{\text{H}}{\underset{|}{\text{C}}}\text{—OH} \rightleftarrows \text{cyclo—CHOH} + \text{acyclic—C=O}$$

The experimental problem is to protest the chloroiridate against reduction to the metal. Where the acyclic alcohol is ethanol, added hydrochloric acid is an effective stabiliser, and cholestan-3-one may be reduced using chloroiridic acid in ethanol by heating for some hours under reflux. Using $(NH_4)_2IrCl_6$ in ethanol, piperidine is an effective stabilising agent, but dimethylsulphoxide, and other sulphoxides, are more effective particularly when a more easily oxidisable alcohol such as isopropanol is used as the hydride source.

Heating *cis*- or *trans*-H $[IrCl_4(Me_2SO)_2]$ in slightly aqueous isopropanol has in fact been shown to lead to the formation of the corresponding iridium hydrides (37) and (38). Acetone is also produced, and $[HIrCl_4(Me_2SO)_2]$ and

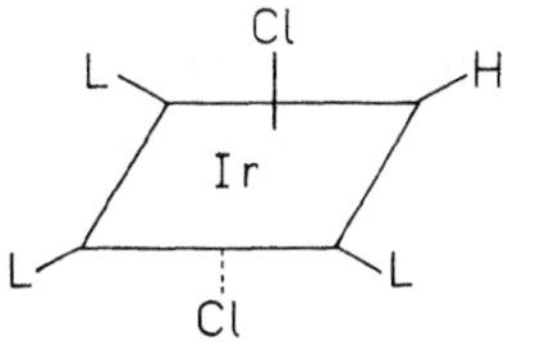

(37)

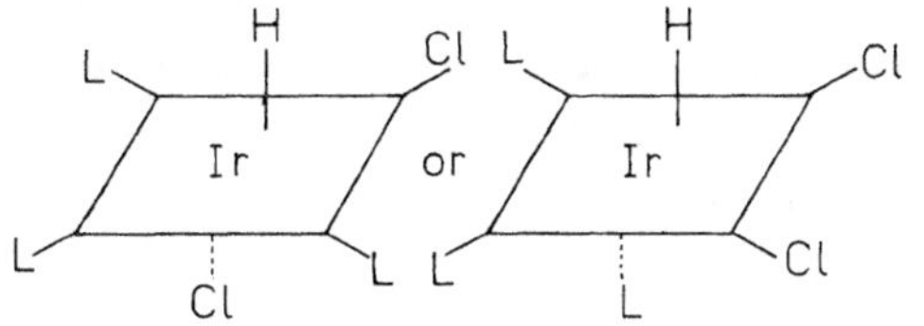

(38)

(L = Me_2SO)

similar complexes act catalytically for hydrogen transfer between isopropanol and cholestan-3-one or 4-*t*-butylcyclohexanone. Addition of a hydride such as (37) or (38) to a carbonyl group:

$$>C{=}O + HIrCl_2L_3 \rightarrow >C(H){-}OIrCl_2L_3$$

followed by reductive regeneration of the hydride can account for the observed catalysis.

This reducing system applied to cholestan-3-one or 4-*t*-butylcyclohexanone is found to lead to a product containing considerably more than the equilibrium proportion of the axial cycloalkanol, but less than is found with the chloroiridic acid – triethylphosphite catalyst discussed above. Thus cholestan-3-one in ethanol or isopropanol with dimethylsulphoxide and M_2IrCl_6, $M = H$ or NH_4, gives a product containing 66–69% of cholestan-3-ol. However, using the chloroiridate in ethanol with HCl or piperidine as stabiliser yields only 30–35% of the axial cholestanol. Thus as may be expected, the stereoselectivity is a function of the nature of the protecting ligand employed.

The ruthenium complex, $RuCl_2(PPh_3)_3$ has also been shown to be effective in hydrogen transfer between secondary alcohols and ketones, [13a] e.g.

$$(PhCH_2)_2CO + PhCH(OH)Me \rightleftarrows (PhCH_2)_2CHOH + PhCOMe$$

4. Decarbonylation of Aldehydes

It was noted above that the carbonyl group appears to be completely inert to hydrogenation in the presence of the $RhCl(PPh_3)_3$ complex. However, with this catalyst aldehydes may undergo decarbonylation, e.g. heptanal with $RhCl(PPh_3)_3$ in benzene at the ordinary temperature yields [14] *n*-heptane, some hept-1-ene, and the carbonyl complex $Rh(CO)(Cl)(PPh_3)_2$. This type of decarbonylation of aldehydes has indeed been generalised as a chemical procedure [15] (cf. Chapter III).

5. Hydrogenation of the Nitro Group

Since the nitro group can be reduced by a variety of reagents, homogeneous catalytic hydrogenation of nitro compounds is of interest rather than of great importance. However, just as with the carbonyl group, there are homogeneous catalysts towards which the nitro group is inert and others which may be used to effect hydrogenation. Thus using $RhCl(PPh_3)_3$ in benzene at 60° the nitro styrenes (39) were reduced only at the olefinic bond [16].

$$\underset{(39)}{ArCH{:}CHNO_2} \rightarrow ArCH_2CH_2NO_2, \quad (Ar = p\text{-}NO_2C_6H_4\text{— or } 3,4\text{-}(MeO)_2C_6H_3\text{—})$$

Similar results were obtained [5] using the $RhCl_3py_3/NaBH_4$ catalyst in dimethylformamide. However, this catalyst system proved to be effective [5] for the hydrogenation of nitrocyclohexane to cyclohexylamine, and for the hydrogenation of a series of aromatic nitro compounds (40)–(42). 4-Dimethylaminonitrosobenzene (43), could also be reduced, but not acetophenone oxime. However, some oximes have been hydrogenated using pentacyanocobaltate (II) as catalyst [18, 19]. This catalyst has also been used for hydrogenation of nitro compounds [19–21], and for reductive amination of ketones [18, 22]. However, with aromatic nitro compounds this catalyst yields mainly the bimolecular reduction products commonly encountered under alkaline conditions.

$PhNO_2 \rightarrow PhNH_2$ (40)

$p\text{-}MeC_6H_4NO_2 \rightarrow p\text{-}MeC_6H_4NH_2$ (41)

$p\text{-}NO_2C_6H_4CO_2H \rightarrow p\text{-}NH_2C_6H_4CO_2H$ (42)

$p\text{-}Me_2NC_6H_4NO \rightarrow p\text{-}Me_2NC_6H_4NH_2$ (43)

Nitrododecane has been hydrogenated at 120° under a pressure of hydrogen to the amine in good yield using $RuCl_2(PPh_3)_3$ as catalyst in oxygen free benzene-ethanol in presence of a base [22a].

Using the $RhCl_3py_3/NaBH_4$/dimethylformamide catalyst the >C=N— bond in benzylidene aniline could be hydrogenated to give benzylaniline, and azobenzene was hydrogenated to hydrazobenzene [17]. The >C=N— bond in various heteroaromatics has also been found to be hydrogenated using this catalyst [5] (cf. Chapter V).

6. The Oxido Group

The oxido group which is known to be cleaved by relatively nucleophilic metal complexes [23] has been reduced catalytically in the presence of $(Co(CN)_5^{3-}$, and also of the $RhCl_3py_3/NaBH_4$ catalyst in dimethylformamide, e.g. (44)→(45) [5, 20].

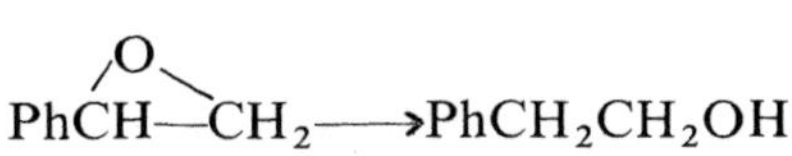

(44) (45)

7. Organic Halides

Many organic halides are well known to react with nucleophilic transition metal complexes including $Co(CN)_5^{3-}$. There is good evidence [24] that displacement of halogen by $Co(CN)_5^{3-}$ is a homolytic process:

$$Rhal + Co(CN)_5^{3-} \rightarrow [Co(CN)_5hal]^{3-} + R\cdot$$

The organic radical may then be reduced in a subsequent step:

$$R\cdot + HCo(CN)_5^{3-} \rightarrow RH + Co(CN)_5^{3-}$$

These reactions are regarded as the basis of the observed hydrogenation of a range of organic halides [25]. Using $Co(CN)_5^{3-}$ in aqueous methanol solution with 1 atm H_2, alkyl iodides, RI, R = Me, Et, *n*-Pr, *i*-Pr, *t*-Bu, *n*-C_5H_{11}, could be hydrogenated to the alkane [25, 26]. However, with $Co(CN)_5^{3-}$ 1,2-dihalides such as $Br(CH_2)_2Br$, $I(CH_2)_2I$, $I(CH_2)_2Cl$ undergo elimination of halogen to give ethylene. 1,4- and 1,5-dihalides behave similarly, and 1,3-dihalides, e.g. $I(CH_2)_3Br$, yield cyclopropane [25, 26]. Allylic halides appear to react mainly by hydrogenolysis of halogen, but the allylic intermediate may add hydrogen at either terminus, e.g. (46) and (47).

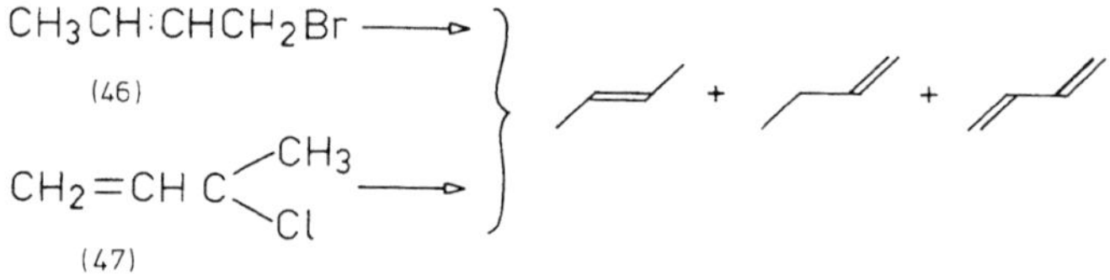

Benzylic halides and α-keto halides such as (48)–(52) undergo hydrogenolysis [27] in the presence of $HCo(CN)_5^{3-}$.

$PhCH_2Br \rightarrow PhCH_3$
(48)

$PhCHBrCH_3 \rightarrow PhC_2H_5$
(49)

p·$BrCH_2C_6H_4CH_2Br \rightarrow$ *p*-$CH_3C_6H_4CH_3$
(50)

$PhCOCH_2Br \rightarrow PhCOCH_3$
(51)

$PhCOCHBrCH_3 \rightarrow PhCOC_2H_5$
(52)

The $RhCl_3py_3/NaBH_4$/dimethylformamide catalyst has also been found to effect catalytic hydrogenolysis of benzylic, allylic, and aryl halides, α-halo esters and ketones, but not alkyl halides [5]. The examples include $PhCH_2Br$, $PhCH_2Cl$, PhCHMeCl, $PhCMe_2Cl$, which gave the corresponding alkyl benzenes, 2-bromo camphor (53), 2-halo-3-oxo steroids (55), (56) and (57), and ethyl bromoacetate, methyl 2-chloro-2-phenyl-ethanoate and methyl 2-chloro-2-phenyl-propanoate.

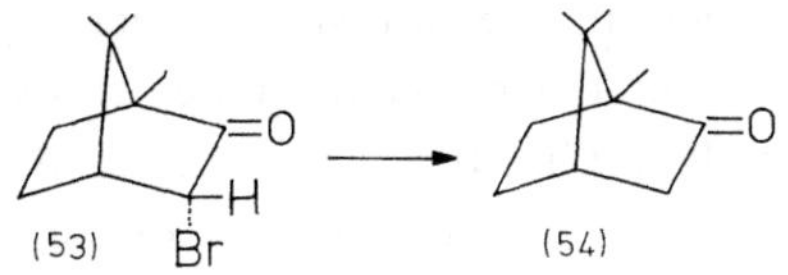

(53) (54)

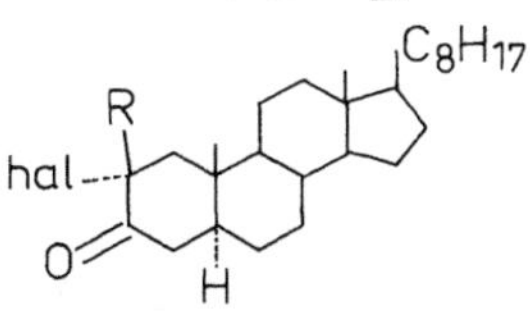

(55) R = H, hal = CL, Br
(56) R = Me, hal = Br

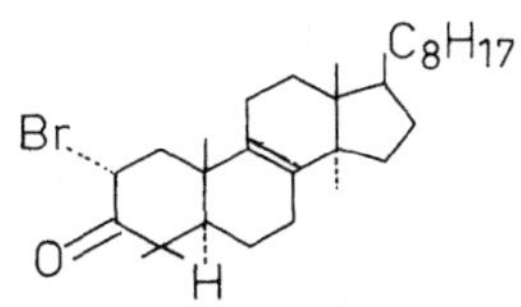

(57)

Me
|
$PhC(Cl)\,CO_2R$
(58)

Me
|
$PhCHCO_2R$
(59)

8. The Stereochemistry of Halogen Hydrogenolysis

Of some interest in this work was the stereochemistry of hydrogenolysis of halides in the presence of the $RhCl_3py_3/NaBH_4$/dimethyl formamide catalyst. The transformation (53)→(54) carried out in the presence of deuterium gave [5] a deuterio-camphor which could be shown to have the stereochemistry (60). The observed halogen displacement with inversion was shown, however, not to indicate a mechanistic preference, but to result from the steric properties of the molecule under study.

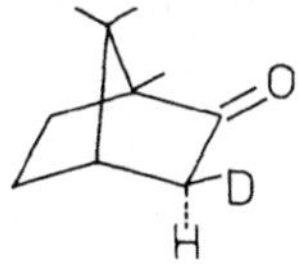

(60)

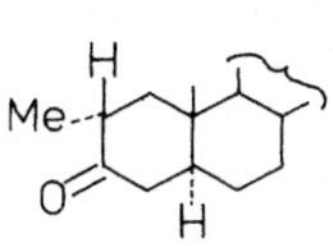

(61)

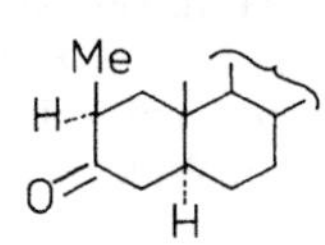

(62)

The steroid (56), on hydrogenation using $RhCl_3py_3/NaBH_4/H_2$, gave [5] a product containing both the 2α-methyl- (61) and 2β-methyl-3-oxo-steroid (62), in a ratio 1:3, indicative of non-stereospecific halogen displacement.

The most informative case, however, was found [5] in the hydrogenolysis of the (−)-form of 2-chloro-2-phenylpropanoic acid and its methyl ester (58, R = H or Me) which gave 2-phenylpropanoic acid or the methyl ester (59, R = H or Me) which proved to be totally racemic. Thus in a particular case hydrogenolysis of halogen may exhibit stereoselectivity, but this is imposed by the steric properties of the molecule under study. The mechanism of halogen displacement clearly has the characteristics of a process involving a radical intermediate, as was noted above for hydrogenolysis by means of $Co(CN)_5^{3-}$ under hydrogen.

References

1. W. L. Diessen and W. L. Groeneveld: *Rec. trav. chim. Pays Bas* **90**, 87 (1971).
2. B. Clarke, M. Green, R. B. L. Osborn, and F. G. A. Stone: *J. Chem. Soc. A*, 167 (1968); J. Browning, C. S. Cundy, M. Green, and F. G. A. Stone: *J. Chem. Soc. A*, 20 (1969).
3. J. Kwiatek and J. K. Seyler: *J. Organomet. Chem.* **3**, 42 (1965); and various patents; cf. *Chem. Abs.* **63**, 2900, 17979 (1965).
4. J. Kwiatek, J. L. Mador, and J. K. Seyler: *J. Amer. Chem. Soc.* **84**, 304 (1962).
5. C. J. Love and F. J. McQuillin: *J. Chem. Soc. Perkin Trans. I*, 2509 (1973).
6. R. R. Schrock and J. A. Osborn: *J. Chem. Soc. Chem. Comm.*, 567 (1970); J. R. Shapley, R. R. Schrock, and J. A. Osborn: *J. Amer. Chem. Soc.* **91**, 2816 (1969).
7. R. S. Coffey: *J. Chem. Soc. Chem. Comm.*, 923 (1967).
8. I. Ojima, M. Nihonyanagi, and Y. Nagai: *J. Chem. Soc. Chem. Comm.*, 938 (1972); I. Ojima, T. Kogure, M. Nihonyanagi, and Y. Nagai: *Bull. Chem. Soc. Japan* **45**, 3506 (1972).
9. I. Ojima, M. Nihonyanagi, and Y. Nagai: *Bull. Chem. Soc. Japan* **45**, 3722 (1972).
10. Y. M. Y. Haddad, H. B. Henbest, J. Husbands, and T. R. B. Mitchell: *Proc. Chem. Soc.*, 361 (1964); cf. E. L. Eliel, T. W. Doyle, R. O. Hutchings, and E. C. Gilbert: *Org. Synth.* **50**, 13 (1970).
11. P. A. Brown and D. N. Kirk: *J. Chem. Soc. (C)*, 1653 (1969); J. C. Orr, M. Mercereau, and A. Sanford: *J. Chem. Soc. Chem. Comm.*, 162 (1970).
12. H. B. Henbest and T. R. B. Mitchell: *J. Chem. Soc. (C)*, 785 (1970).
13. Y. M. Y. Haddad, H. B. Henbest, J. Husbands, T. R. B. Mitchell, and J. Trocha-Grimshaw: *J. Chem. Soc. Perkin I Trans.*, 592, 596 (1974).

13a. Y. Sasson, P. Albin, and J. Blum: *Tetrahedron Letters*, 833 (1974).

14. F. H. Jardine and G. Wilkinson: *J. Chem. Soc. (C)*, 270 (1967).
15. J. Tsuji and K. Ohno: *Synthesis* **1**, 157 (1969); *idem.*, *Tetrahedron Letters*, 3969 (1965); *J. Amer. Chem. Soc.* **88**, 3452 (1966); **90**, 99 (1968).

16. R. E. Harmon, J. L. Parsons, D. W. Cooke, S. K. Gupta, and J. Schoolenberg: *J. Org. Chem.* **34**, 3684 (1969).
17. P. Abley, I. Jardine, and F. J. McQuillin: *J. Chem. Soc. (C)*, 840 (1971).
18. M. Murakami and J. Kang: *Bull. Chem. Soc. Japan* **36**, 763 (1963).
19. M. Murakami, R. Kawai, and K. Suzuki: *J. Chem. Soc. Japan* **84**, 669 (1963).
20. J. Kwiatek, I. L. Mador, and J. K. Seyler: *J. Amer. Chem. Soc.* **84**, 304 (1962).
21. J. Kwiatek, I. L. Mador, and J. K. Seyler: *Adv. in Chem. Series Amer. Chem. Soc.* **73**, 201 (1963).
22. M. Murakami, K. Suzuki, M. Fujishigi, and J. Kang: *J. Chem. Soc. Japan* **85**, 235 (1964).
22a. J. F. Knifton: *J. Org. Chem.* **40**, 519 (1975).
23. F. R. Jensen, V. Mador, and D. H. Buchanan: *J. Amer. Chem. Soc.* **92**, 1414 (1970).
24. J. Halpern and J. P. Maher: *J. Amer. Chem. Soc.* **86**, 2311 (1964); P. E. Chock and J. Halpern: *ibid.* **91**, 582 (1969); J. Halpern and P. E. Phelan: *ibid.* **94**, 1881 (1972).
25. J. Kwiatek and J. K. Seyler: *Adv. in Chem. Series Amer. Chem. Soc.* **70**, 207 (1968).
26. J. Kwiatek and J. K. Seyler: *J. Organomet. Chem.* **3**, 42 (1965); P. B. Chock and J. Halpern: *J. Amer. Chem. Soc.* **91**, 582 (1969).
27. J. Kwiatek: *Catal. Rev.* **1**, 37 (1967).

CHAPTER VII

ISOMERISATION AND SPECIFIC DEUTERIATION

1. Isomerisation

It was noted in Chapter I that in hydrogenation the process of addition of a metal hydride across an alkene bond is fundamentally reversible, i.e.:

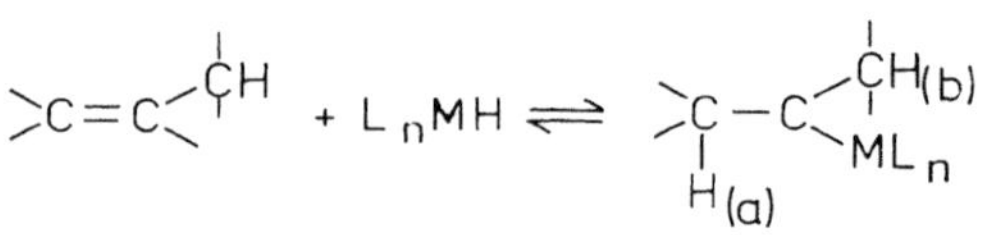

(1) (2)

Thus the mechanism of hydrogenation contains the possibility of alkene isomerisation since it is also evident that the reverse process of elimination of L_nMH from the metal alkyl (2) may involve either hydrogen $H_{(a)}$, leading back to the parent alkene (1), or $H_{(b)}$ yielding an isomerised alkene.

Competing with the elimination reaction is the hydrogenolysis of the metal alkyl (2) to give the alkane and hence the critical factor determining the extent of isomerisation is the rate of elimination relative to the rate of hydrogenolysis of the metal alkyl intermediate. An important second factor, however, derives from the way in which the homogeneous hydrogenation catalysts discriminate between more and less substituted alkene bonds. Thus in a particular case such as (3), the isomer (4) may be formed by isomerisation

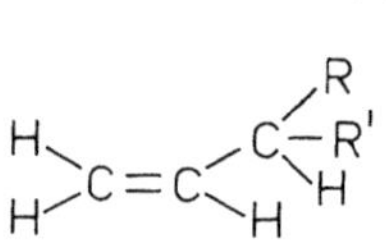

(3)

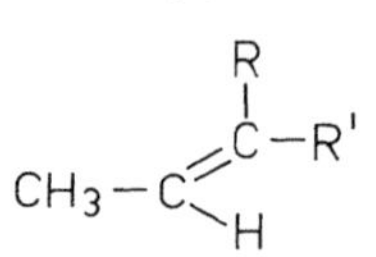

(4)

and accumulate when the substituents R and R′ stabilise this isomer against co-ordination to the catalyst and hence hydrogen addition.

It is also clear that the mechanistic pattern of addition and elimination of a metal hydride may lead to *cis-trans* stereomutation of an alkene. Thus in

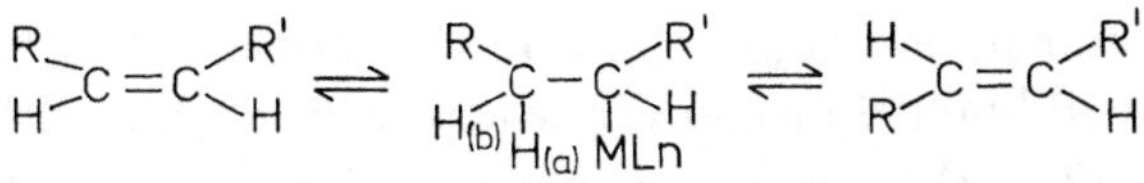

(5) (6) (7)

a metal alkyl intermediate (6), formed by addition of HML_n to (5), elimination may involve $H_{(a)}$ which leads back to the *cis*-alkene (5), or elimination of hydrogen $H_{(b)}$ which yields the *trans*-isomer (7). Since *trans*-alkenes are very generally less strongly co-ordinated, the *trans*-isomer may for a time accumulate during hydrogenation.

Thus the incidence of isomerisation during hydrogenation is necessarily a function of the structure of the alkene as well as of the structure and reactivity of the metal complex catalyst. It has been found, however, that incidental experimental factors such as the solvent used, and particularly the presence or absence of oxygen may also be very important in determining the extent of isomerisation.

Initial studies using the chloro*tris* (triphenylphosphine) rhodium (I) catalyst suggested a total absence of isomerisation or stereomutation in hydrogenation with this complex [1]. Closer study, however, revealed that this conclusion is not entirely correct.

In the case of the relatively simple alkene, *cis*-pent-2-ene (8) it was found [2] that hydrogenation in benzene-ethanol (1:1) using $RhCl(PPh_3)_3$ leads to formation of substantial amounts of *trans*-pent-2-ene (10) and to a lesser quantity also of pent-1-ene (11) during the reaction. These products may be

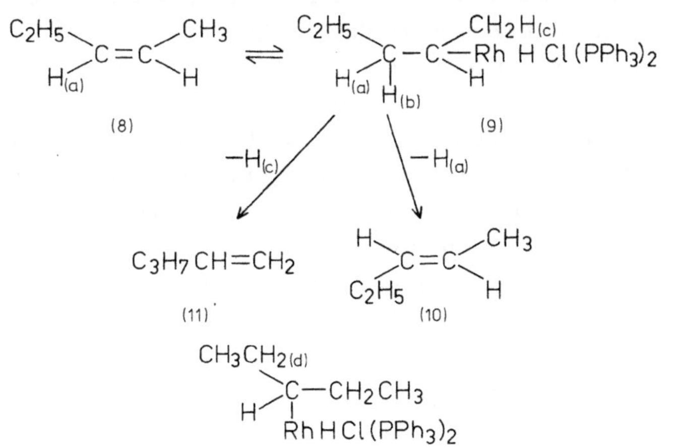

(12)

explained by elimination of $H_{(a)}$ or $H_{(c)}$ from an intermediate (9). It will also be recognised that addition of metal hydride across the alkene bond in (8) may lead to adducts (9) or (12), and that elimination of one of the hydrogens $H_{(d)}$ from (12) may also lead to *cis*- or *trans*-pent-2-ene.

From comparative studies with the group of complexes, Rh(hal) $(PPh_3)_3$, hal = Cl, Br, I, there was an indication of some reduction in isomerisation with the larger halogens [2], i.e. Br on I, but the effect is small.

Isomerisation has been noted [3] during the hydrogenation of 4-*t*-butyl-methylenecyclohexane (13) in the presence of $RhCl(PPh_3)_3$ which gives in substantial amount the less easily hydrogenated isomer, 4-*t*-butylmethyl-cyclo hex-1-ene (14). The degree of isomerisation also appeared to be some-

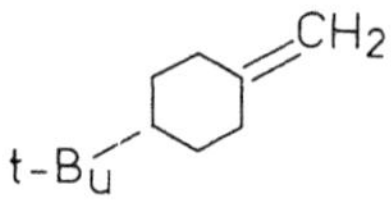

(13)

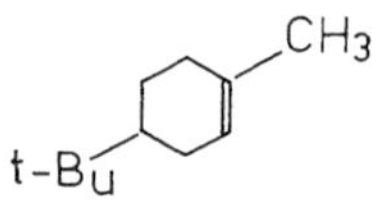

(14)

what greater when 95% ethanol rather than benzene ethanol (1:1) was used as solvent [3], and smaller in presence of a higher pressure of hydrogen. It was subsequently [4] shown that isomerisation is in fact minimal if the $RhCl(PPh_3)_3$ complex is pre-saturated with hydrogen so that the dihydride, $RhH_2Cl(PPh_3)_3$ is initially present in high concentration, and that in benzene solvent, in particular, isomerisation is largely suppressed. The presence of oxygen, on the other hand, appears to promote olefin isomerisation especially if the solvent contains ethanol [4].

There is evidence of reaction of $RhCl(PPh_3)_3$ with oxygen to give a peroxy derivative as well as formation of the phosphine oxide, Ph_3PO. In addition there is evidence of oxidation of the ethanol solvent leading to formation of a rhodium carbonyl complex. The nature of these processes has not been fully clarified [5], but the presence of oxygen in the system is clearly disadvantageous in view of the sensitivity of the $RhCl(PPh_3)_3$ complex.

This effect of oxygen in promoting isomerisation applies also to hydrogenations carried out using the ruthenium catalyst, $RuCl_2(PPh_3)_3$. This complex shows some activity for alkene isomerisation in benzene solution [6], but in the presence either of oxygen, added peroxides, or of peroxides present in the alkene it is highly active for isomerisation [7]. These results, however, refer to reaction at 100° which does not correspond to the circumstances of catalytic hydrogenation, and although $RuCl_2(PPh_3)_3$ or derived hydrido complexes do effect alkene isomerisation under mild conditions [8], in hydrogenation isomerisation is probably minimal [8].

It is important to recognise, however, that in many recorded studies isomerisation has not always been looked for, and that at times isomerisation may not be directly detectable. An interesting example [9] arises from reduction of cyclo-octene at $RhCl(PPh_3)_3$ in deuterium using benzene as solvent. The ^2H-cyclo-octane contained only 87.8% of the $^2H_2$-species with 5.4% $^2H_1$, 3.8% $^2H_0$, 2.8% $^2H_3$ and smaller amounts of products up to $^2H_9$-cyclo-octane. By contrast cyclohexene gave a product comprising 97% $^2H_2$-cyclohexane, i.e. the extent of isomerisation and exchange is a function of ring size, an aspect more fully considered below under deuteriation. There is also no doubt that these results arise from an addition-elimination sequence since at incomplete reduction the recovered cycloalkene contained deuterium. Similar results were recorded from a study of catalysed tritiation of cyclohexene [10] with, however, the interesting indication that exchange is relatively less important if hydrogenation is carried out at a somewhat elevated temperature, viz. 60° instead of 30 °C.

The example which first drew particular attention to the importance of isomerisation during homogeneous hydrogenation was encountered in the attempted reduction of damsin (15). This sesquiterpene lactone in the presence of $RhCl(PPh_3)_3$ was found [11] to undergo preferential isomerisation to iso-damsin (16), rather than hydrogen addition. This example illustrates very well the principle of migration of the olefinic bond from a less

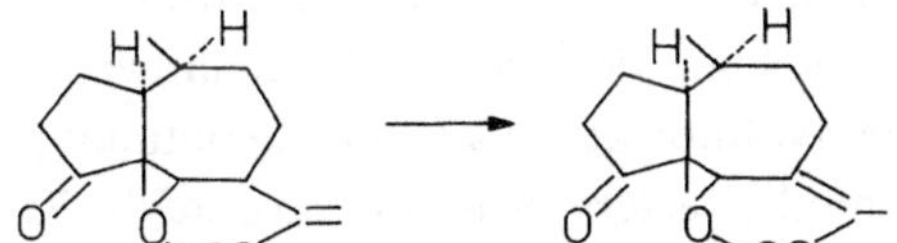

(15) (16)

to a more stable position. However, the importance of small structural factors in determining the rate of hydrogenation *vs* isomerisation is emphasised by the behaviour of two structurally closely related materials. Psilostachyine (17) and confertiflorin (18), under the same conditions, give the expected dihydro-derivative without isomerisation.

Experience of hydroformylation [12] with the complex $RhH(CO)(PPh_3)_3$

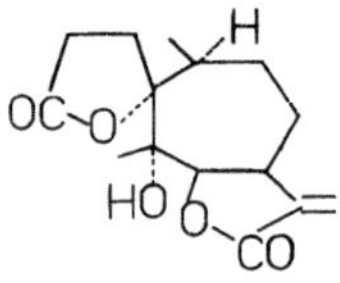

(17)

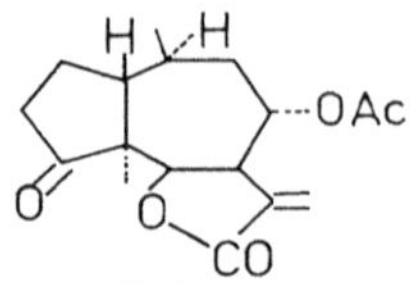

(18)

indicated that this catalyst is quite active in the isomerisation of alkenes, and a fuller study showed [13] that with $RhH(CO)(PPh_3)_3$ in benzene isomerisation of pent-1-ene occurs at a rate similar to the rate of hydrogenation. As is frequently observed, the initial product of isomerisation is the *cis*-isomer of pent-2-ene, which more slowly undergoes stereomutation to *trans*-pent-2-ene.

The complex $IrCl(PPh_3)_3$ appears to be more active in hydrogenation than its rhodium analogue but is also particularly effective for olefin isomerisation [14].

Thus alkene isomerisation is to a greater or lesser extent a general property of complexes which are active in homogeneous hydrogenation.

2. Isomerisation with Naturally Occurring Unsaturated Fatty Acids

Alkene isomerisation is of particular concern in the hydrogenation of the naturally occurring unconjugated polyunsaturated fatty acid esters. The controlled hydrogenation of esters of linoleic and linolenic acid has been the subject of considerable study using metal carbonyl catalysts such as $Fe(CO)_5$, Pt(II) chloride systems assisted by $SnCl_2$, and also using $RhCl_3py_3$ or $NiCl_2$ activated by sodium borohydride in dimethylformamide. Of these catalysts, $NiCl_2/NaBH_4$ in dimethylformamide [15] has been shown to effect small scale hydrogenation of methyl linoleate to a product containing 15% methyl stearate and 85% of the methyl esters of monoenoic acids. This result indicates marked preferential hydrogenation of the dienoic in comparison with the monoenoic ester, but significantly there was an apparent absence of translocation or stereomutation of the alkene bonds. The monoenoic ester product by mass spectral and n.m.r. analysis [15, 16] showed a high proportion of *cis*-mono olefin product with the alkene bond in the positions occupied in the parent dienoic ester.

By contrast hydrogenation of methyl linoleate using the $RhCl_3py_3/NaBH_4/dmf$ catalyst [15] gave a monoenoic ester product showing extensive translocation of alkene bonds, and a high proportion of material isomerised to the *trans* olefin geometry.

This indication that a nickel based catalyst is relatively less active in isomerisation is borne out in other examples. For oct-1-ene with a series of complexes $(Ph_3P)_2MX_2$, M = Pt, Pd, Ni, X = hal, it was found [17] that the nickel complexes, X = Br, I, cause very little alkene isomerisation whereas the related platinum and palladium complexes are rather effective in this respect. This group of complexes $(Ph_3P)_2Mhal_2$ are not, however, particularly active as catalyst for hydrogenation [17]. Addition of stannous chloride which replaces a halide by a $[Snhal_3]^-$ ligand, however, enhances the hydrogenation rate, and complexes of the type L_2PtX_2, L = e.g. Ph_3P, X = hal, in presence of $SnCl_2$ have been applied [18] to the hydrogenation of methyl linoleate and similar unsaturated esters in benzene-methanol solution at 90 °C and 40 atm H_2. However, as noted above [16], hydrogenation is accompanied by extensive *cis-trans* isomerisation [18].

Various transition metal acetylacetonates such as $Ni(acac)_2$ or $Co(acac)_3$ which have also been used [19] for hydrogenation of poly-unsaturated fatty esters also appear to promote a good deal of *cis-trans* alkene isomerisation. The mode of action of these acetylacetonates is not known, but the methanol solvent appears to play an important part, perhaps as a hydride source.

The carbonyls $Fe(CO)_5$ and $Co_2(CO)_8$ have also been employed [20] for hydrogenation of fatty acid esters, but only at a relatively high pressure of hydrogen, ca. 30 atm H_2, and at ca. 180 °C. However, with these catalysts also there is considerable translocation of the alkene bonds, and the monoenoic ester obtained from e.g. methyl linoleate is found to contain a considerable percentage of the trans isomer.

It is evident that outside the range of the small number of highly active catalytic complexes, and notably $RhCl(PPh_3)_3$, isomerisation and translocation of olefinic bonds is a common phenomenon, and that the best catalysts such as $RhCl(PPh_3)_3$ are not entirely free of this complication.

3. Deuteriation

Specific deuteriation of maleic (19) and fumaric acids (20) to meso- and dl-1,2-^{2}H-succinic acids respectively was the key observation which established [1] *cis*-1,2-addition as the mode of catalysis by $RhCl(PPh_3)_3$. This specific deuteriation of dimethyl fumarate has been employed as the first step in a synthesis of dl-2,3-$^2H_2$-1,4-dibromobutane [20a].

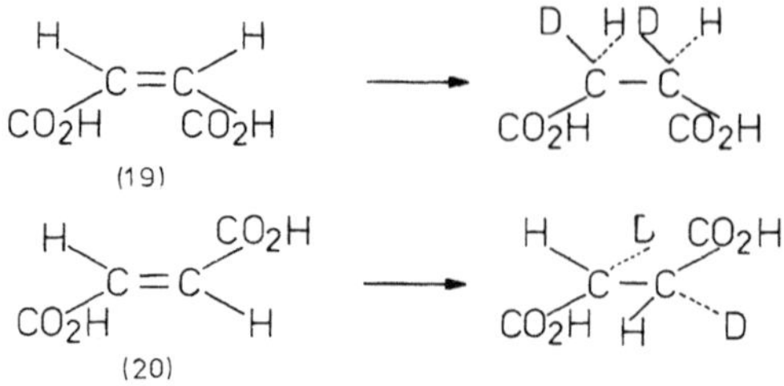

Similar observations have been made with the $RhCl_3py_3/NaBH_4$/dimethylformamide catalyst system [21], although it may be noted that with this catalyst diphenylacetylene is hydrogenated to *trans*-stilbene [21].

However, clear evidence that the catalytic process does not invariably involve simple *cis*-addition of hydrogen comes from the results [9] noted above for deuteriation of cyclohexene and cyclo-octene. With $RhCl(PPh_3)_3$ in benzene, cyclohexene gave 97% $^2H_2$-cyclohexane whereas cyclo-octene under the same conditions gave a product showing 87.7% $^2H_2$- and appreciable amounts of $^2H_0$-, $^2H_1$- and $^2H_3$-cyclo-octane.

Similar studies [22] using the $RhCl_3py_3/NaBD_4$/dimethylformamide catalyst also showed extensive scrambling of hydrogen and deuterium in the ^2H-cycloalkane products from both cyclohexene and cyclo-octene. By contrast the corresponding $NiCl_2/NaBH_4$ system using nickel as the transition metal, catalysed deuteriation of methyl sorbate (21) to give, in good yield, the 4,5-$^2H_2$-pent-2-enoic ester (22), as indicated by n.m.r. evidence [23].

(21) (22)

The evident importance of the catalyst in minimising isomerisation is illustrated by a number of examples using the $RhCl(PPh_3)_3$ catalyst. For example, pent-2-enoic acid (23) with $RhCl(PPh_3)_3$ in H_2/HT, gave a pentanoic acid showing tritium incorporation at C-2, C-3, C-4, and C-5 of 50:48:1:0.5, indicating very little olefinic bond migration [24].

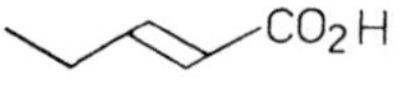

(23)

Chloro*tris* (triphenylphosphine) rhodium (I) was also successfully applied to preparing *cis*-1,2-$^2H_2$-cyclo-octa-3,5,7-triene (25), required for an n.m.r. study of the thermodynamics of ring inversion of cyclo-octatrienes [25].

(24) (25)

Specific *cis*-1,2-deuteriation was also observed in the preparation of erythro-1,2-$^2H_2$-1-phenyl-2-methoxyethane ($PhCHDCHDOCH_3$) from *cis*-β-methoxystyrene (*cis*-$PhCH{=}CHOCH_3$) using $RhCl(PPh_3)_3$ as catalyst in deuterium [26].

In another similar example the $RhCl(PPh_3)_3$ catalyst in benzene was used [27] to effect *cis*-deuteriation of the isomeric 2,5-dimethoxydihydrofuran derivatives (26) and (27).

(26)

(27)

Many of these examples, however, are not a very critical test of specificity since there is little opportunity for isomerisation in the molecules studied. More critical evidence is provided by the apparently specific deuteriation [28] of methyl oleate (28) and methyl linoleate (29), and of a series of terminal and in-chain higher alkenes [29] using $RhCl(PPh_3)_3$ as catalyst.

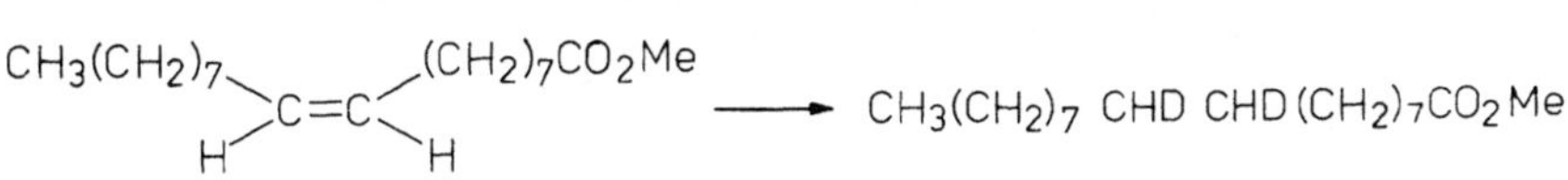

(28)

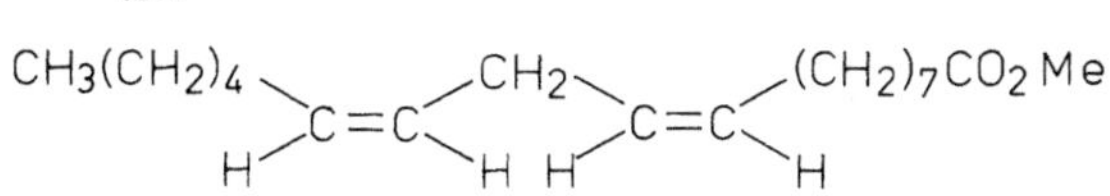

(29)

As was noted above, cyclohexene gives $^2H_2$-cyclohexane of satisfactory isotopic purity by use of $RhCl(PPh_3)_3$ in presence of deuterium, and a number of substances containing a cyclohexene moiety have been converted into the $^2H_2$-derivative successfully using this catalyst. The 5-6-$^2H_2$-hydrindan-1-one (30) required for studies of mass spectral breakdown was prepared in this way, ^2H-content: $^2H_2$ 82, $^2H_1$ 18%. The analogue (31), prepared similarly, also showed a closely similar deuterium incorporation, viz.: $^2H_2$ 83, $^2H_1$ 13, $^2H_0$ 4%. It is clear from the deuterium content in these

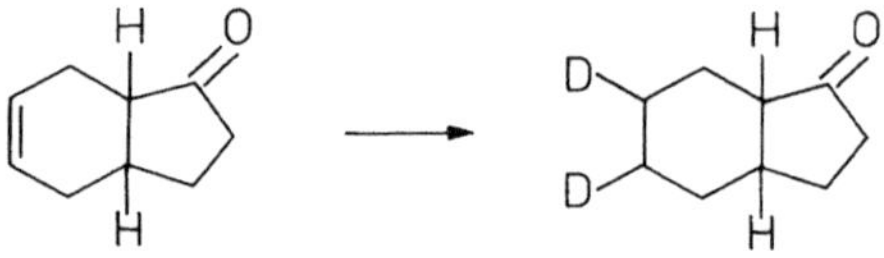

(30)

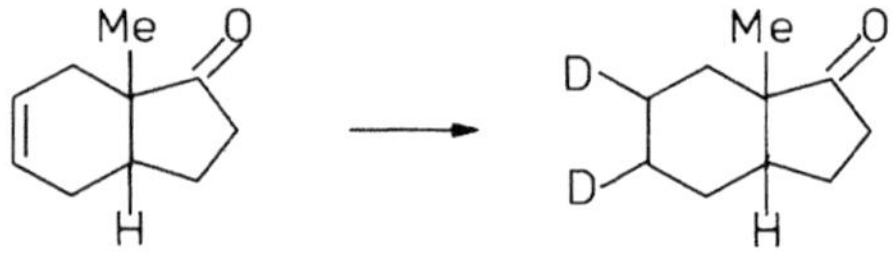

(31)

examples, however, that in deuterium addition catalysed by $RhCl(PPh_3)_3$ in CH_3OD solution deuterium-hydrogen exchange is not completely eliminated.

4. Stereoselectivity in Deuteriation

Deuteriation of the steroid derivative androsta-1,4-dien-3,17-dione, (32), showed [31] the expected selectivity in discrimination between the 1,2- and 4,5-alkene centres, but also high stereoselectivity in deuteriation. The product

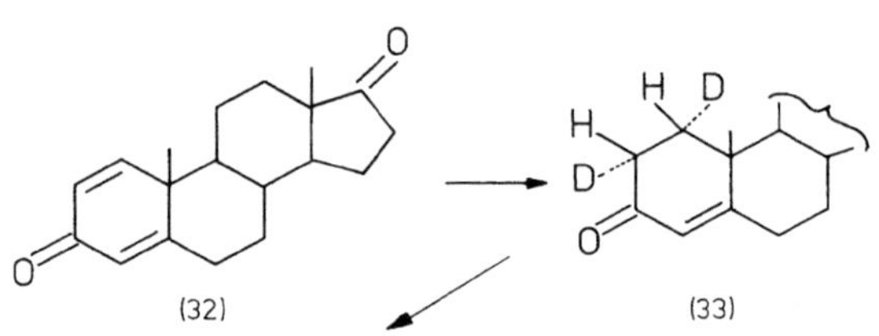

(32) (33)

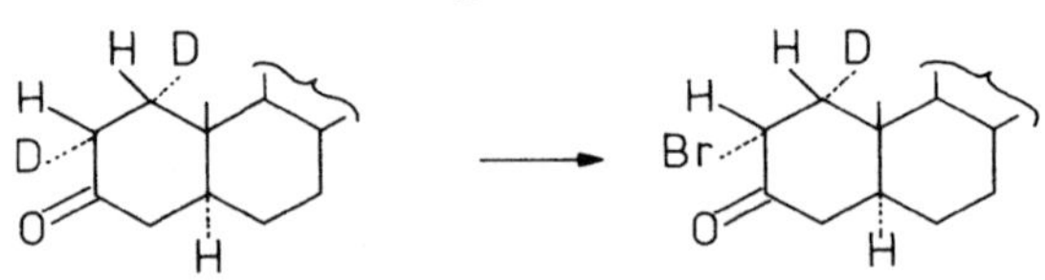

(34) (35)

(33), (85% $^2H_2$), could be shown to be the 1α, 2α-$^2H_2$-steroid (33), since the 2-proton in the 2α-bromo derivative (35), formed by reduction to (34) and bromination, which are processes of known stereochemistry, showed a $J_{H,H}$ value of 6.5. Hz. The selective reactivity of the 1,2-alkene bond revealed by these studies has also been successfully employed [31a] to obtain specifically tritiated [1,2-$^3H_2$] ergosterone and thence [1-^{3}H] ergosterol and [1-^{3}H] ergocalciferol.

5. The Effects of the Solvent

Further studies have indicated that reasonably specific catalytic deuteriation without ^{1}H—^{2}H- exchange is dependent on the use of a relatively non-protic solvent medium. This is illustrated by the following results for deuteriation of cholest-2-ene (36) using $RhCl(PPh_3)_3 + D_2$ in different solvents [32].

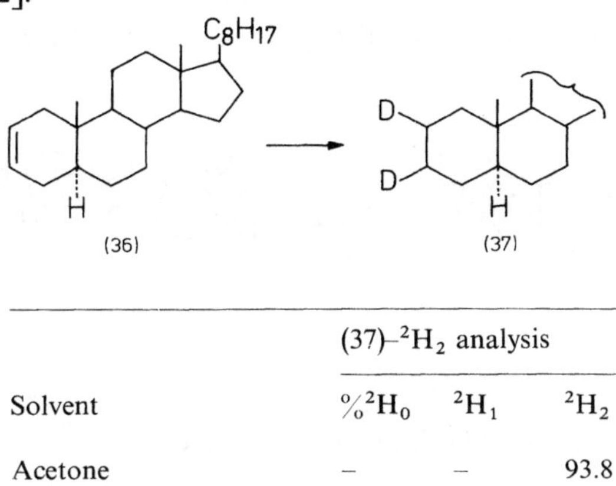

(36) (37)

Solvent	(37)–$^2H_2$ analysis			
	%$^2H_0$	$^2H_1$	$^2H_2$	$^2H_3$
Acetone	–	–	93.8	5.2
Tetrahydrofuran	–	–	94.7	4.8
Benzene	–	–	93.6	5.3
Benzene: Ethanol (1:1)	1.2	2.6	91.9	3.0
Benzene: Ethanol (1:4)	1.0	16.2	77.4	3.7
Benzene: Methanol (3:7)	9.7	38.8	48.5	2.2
Benzene: CH_3OD	–	–	93.6	6.7

It is evident that use of methanol or ethanol as solvent is, not surprisingly, a source of ^{1}H—^{2}H exchange, and incomplete deuteriation of the product.

Dioxan is a useful solvent for specific deuteriation. The steroid derivative (38), for example, with $RhCl(PPh_3)_3$ and D_2 in dioxan gave the 6,7-$^2H_2$ product (39) in satisfactory isotopic purity [33]. This also constitutes an

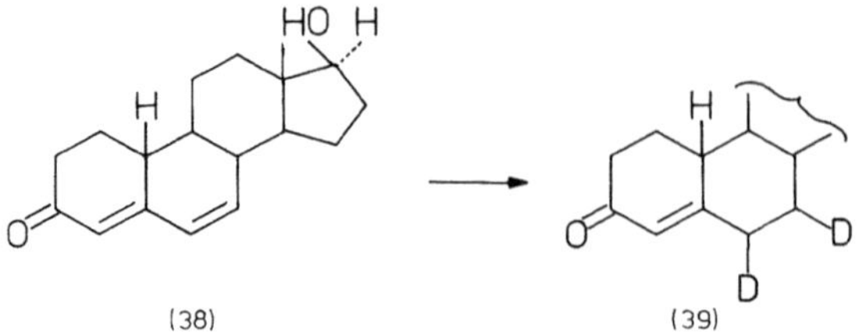

(38) (39)

example of selective reduction as between the 6,7- and the more substituted 4,5-alkene centre, and an example of stereospecific deuterium addition. The product (39) could be shown to be the $6\beta,7\beta$-$^2H_2$-derivative [33].

A further example, using acetone as solvent, also illustrates the specific catalytic deuteriation of the side chain of the steroidal sapogenin [34]. Thus (40), derived from the parent sapogenin by bromination and debromination, gave with $RhCl(PPh_3)_3$ and D_2 in acetone a product containing 97% of the $^2H_2$-derivative (41), the stereochemistry of which could be established by n.m.r. [34].

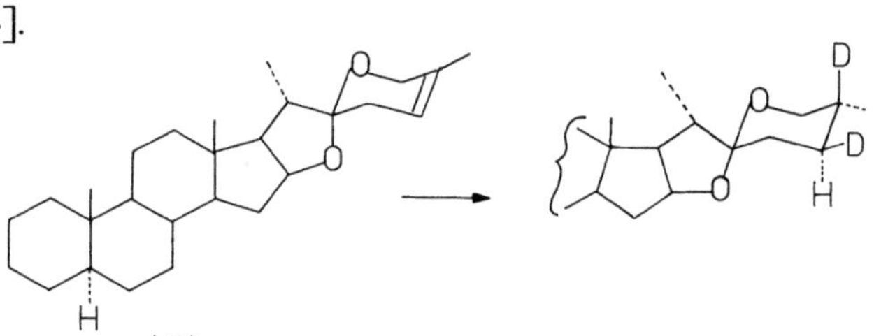

(40) (41)

6. The Effect of Ring Size

It was noted above that the extent of deuterium scrambling during deuteriation of a cycloalkene at $RhCl(PPh_3)_3$ is a function of the ring size. For a cyclohexene, scrambling and ^1H—^2H exchange is minimal, but greater than in cyclopentene. However, medium sized cycloalkenes up to cyclodecene show extensive deuterium scrambling which falls again in cyclododecene or cyclotridecene [35]. Representative data are given in the table for a series of *cis*-cycloalkenes with $RhCl(PPh_3)_3$ and D_2 in benzene:

Ring size	5	6	7	8	9	10	12	13
% $^2H_2$	96.4	82.9	84.2	48.0	42.9	28.7	81.0	82.0

It is significant that the rate of hydrogenation of cycloalkenes falls with increasing ring size between cyclopentene and *cis*-cyclodecene, and then rises somewhat for cyclododecene [35]. Equally, it may be noted that for cycloalkyl derivatives the rate of *cis*-elimination processes shows a marked increase with ring size for rings between six and ten membered [36]. Both factors will contribute to assisting a process:

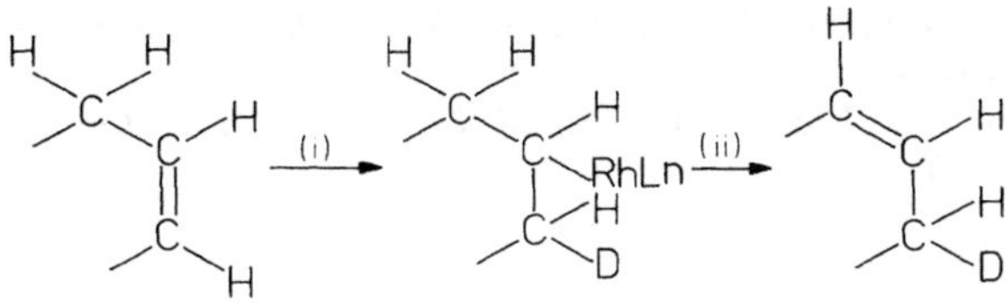

leading to isomerisation and deuterium scrambling, i.e. as step (i) is slowed and step (ii) increased in rate.

7. 1,4-Deuteriation of 1,3-Dienes

Specific 1,4-deuteriation of a 1,3-diene system has been observed [37] in the reaction of methyl sorbate (42) with deuterium catalysed by the chromium carbonyl catalyst $Cr(CO)_3(C_6H_5COOMe)$. The product showed $^2H_2$ 94, $^2H_3$ 2.1 and $^2H_0$ 3.0% and gave ^1H- and ^2H-n.m.r. spectra consistent with structure (43).

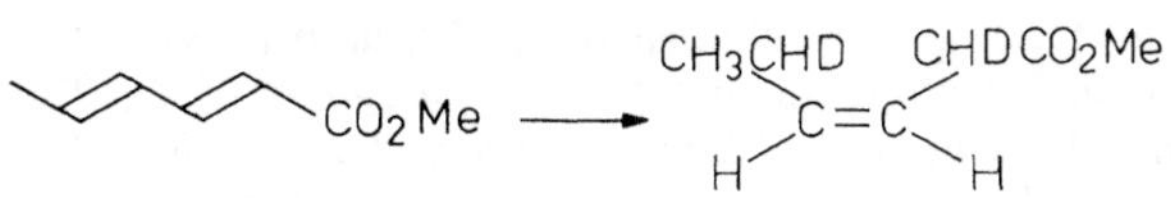

(42) (43)

References

1. J. A. Osborn, F. H. Jardine, J. F. Young, and G. Wilkinson: *J. Chem. Soc. (A)*, 1711 (1966); F. H. Jardine, J. A. Osborn, and G. Wilkinson: *J. Chem. Soc. (A)*, 1574 (1967); A. J. Birch and K. A. M. Walker: *J. Chem. Soc. (C)*, 1894 (1966).
2. G. C. Bond and R. A. Hillyard: *Disc. Faraday Soc.* **46**, 20 (1968).
3. R. L. Augustine and J. van Peppen: *Ann. New York Acad. Sci.* **158**, 482 (1969).
4. R. L. Augustine and J. van Peppen: *J. Chem. Soc. Chem. Comm.*, 495 (1970).
5. *Idem.*, *ibid.*, 497.
6. B. Hudson, P. C. Taylor, D. E. Webster, and P. B. Wells: *Discuss. Faraday Soc.* **46**, 37 (1968); D. F. Ewing, B. Hudson, D. E. Webster, and P. B. Wells: *J. Chem. Soc. Dalton Trans.*, 1287 (1972).

7. J. E. Lyons: *J. Chem. Soc. Chem. Comm.*, 562 (1971); *J. Org. Chem.* **36**, 2497 (1971).
8. P. S. Hallman, B. R. McGarvey, and G. Wilkinson: *J. Chem. Soc. (A)*, 3143 (1968); D. Rose, J. D. Gilbert, R. P. Richardson, and G. Wilkinson: *ibid.*, 2610 (1969).
9. J. G. Atkinson and M. O. Luke: *Canadian J. Chem.* **48**, 3580 (1970).
10. A. L. Odell, J. B. Richardson, and W. R. Roper: *J. Catal.* **8**, 393 (1967).
11. J. F. Biellman and M. J. Jung: *J. Amer. Chem. Soc.* **90**, 1673 (1968).
12. D. Evans, J. A. Osborn, and G. Wilkinson: *J. Chem. Soc. (A)*, 3133 (1968).
13. M. Yagupsky and G. Wilkinson: *J. Chem. Soc. (A)*, 941 (1970).
14. H. van Gaal, H. G. A. M. Cuppers, and A. van der Ent: *J. Chem. Soc. Chem. Comm.*, 1694 (1970).
15. P. Abley and F. J. McQuillin: *J. Catal.* **24**, 536 (1972); A. G. Hinze and D. J. Frost: *ibid.*, 540.
16. D. E. Minnikin, P. Abley, F. J. McQuillin, K. Kusamran, K. Maskens, and N. Polgar: *Lipids* **9**, 135 (1974).
17. P. Abley and F. J. McQuillin: *Discuss. Faraday Soc.* **46**, 31 (1968).
18. J. C. Bailar and H. Itatani: *J. Amer. Chem. Soc.* **89**, 1592 (1967); H. Itatani and J. C. Bailar: *ibid.* **89**, 1600 (1967).
19. E. A. Emken, E. N. Frankel, and R. O. Butterfield: *J. Amer. Oil Chem. Soc.* **43**, 14 (1966).
20. E. N. Frankel, E. A. Emken, H. M. Peters, V. L. Davison, and R. O. Butterfield: *J. Org. Chem.* **29**, 3292 (1964); E. N. Frankel, E. A. Emken, and V. L. Davison: *J. Amer. Oil Chem. Soc.* **43**, 307 (1966); E. N. Frankel and S. Metlin: *ibid.* **44**, 37 (1967); E. N. Frankel, G. P. Jones, V. L. Davison, E. Emken, and H. J. Dutton: *ibid.* **42**, 130 (1965).
20a. R. H. Grubbs and T. K. Brunck: *J. Amer. Chem. Soc.* **94**, 2538 (1972).
21. P. Abley and F. J. McQuillin: *J. Chem. Soc. Chem. Comm.*, 1503 (1969).
22. C. J. Love: Ph.D. Thesis, University of Newcastle upon Tyne, 1972, p. 38.
23. A. G. Hinze and D. J. Frost: *J. Catal.* **24**, 540 (1972).
24. H. Simon and O. Berngruber: *Tetrahedron*, (1970) **26**, 161; *idem.*, *ibid.*, 1401.
25. M. St. Jaques and R. Prud'homme: *Tetrahedron Letters*, 4833 (1970).
26. D. Slack and M. C. Baird: *J. Chem. Soc. Chem. Comm.*, 701 (1974).
27. D. Gaghaire and P. Voltero: *Bull. Soc. chim. France*, 164 (1970).
28. A. J. Birch and K. A. M. Walker: *J. Chem. Soc. (C)*, 1894 (1966).
29. J. R. Morandi and H. B. Jensen: *J. Org. Chem.* **34**, 1889 (1969).
30. B. Zech, G. Jones, and C. Djerassi: *Chem. Ber.* **100**, 3204 (1967).
31. C. Djerassi and J. Gutzwiller: *J. Amer. Chem. Soc.* **88**, 4537 (1966); cf. H. J. Brodie, K. J. Kripalani, and G. Possanza: *ibid.* **91**, 1241 (1969).
31a. B. Pelc and E. Kodicek: *J. Chem. Soc. (C)*, 3415 (1971); cf. D. E. M. Lawson, B. Pelc, P. A. Bell, P. W. Wilson, and E. Kodicek: *Biochem. J.* **121**, 673 (1971).
32. W. Voelter and C. Djerassi: *Chem. Ber.* **101**, 58 (1968).
33. H. J. Brodie, C. E. Hay, and T. A. Willstruck: *J. Org. Chem.* **37**, 3361 (1972).
34. W. H. Fane, A. Failli, and C. Djerassi: *J. Org. Chem.* **35**, 2571 (1970).
35. B. Bayerl, M. Wahren, and J. Graefe: *Tetrahedron* **29**, 1837 (1973).
36. J. Sicher: *Angew. Chem. Internat. Edn.* **11**, 200 (1972).
37. E. N. Frankel, E. Selke, and C. A. Glass: *J. Amer. Chem. Soc.* **90**, 2446 (1968).

CHAPTER VIII

LIGANDS AND SOLVENTS

Transition metal hydrides which are active in hydrogen transfer and catalysis, e.g. (1)–(9), very generally carry stabilising ligands such as

$[HCo(CN)_5]^{3-}$	$RhCl(H_2)(PPh_3)_2$	$RhH(CO)(PPh_3)_3$
(1)	(2)	(3)
$RuHCl(PPh_3)_3$	$RuH(OCOR)(PPh_3)_3$	$PtHCl(PPh_3)_2$
(4)	(5)	(6)
$CoH(CO)_4$	$MnH(CO)_5$	$Mo(cp)H_2$
(7)	(8)	(9)

a phosphine, carbonyl, cyanide or π-cyclopentadienyl, capable of retrodative π-bonding. Ligands of this type are, however, not altogether essential. The pentamine hydridorhodium derivative (10) for example, is known as stable salts [1], and the hydride reacts readily with ethylene, propene or but-1-ene to give adducts (11), (R = Et, Pr, Bu), which have been characterised [1].

$[RhH(NH_3)_5]^{2+}$ + 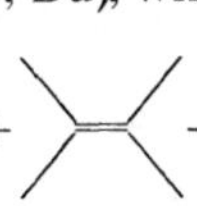 $\rightarrow [Rh(R)(NH_3)_5]^{2+}$

(10) (11)

Whilst π-acceptor ligands are clearly not essential for the existence or typical alkene addition reaction of the transition metal hydrides, ligands such as PR_3, or CO do contribute to the stability of the metal hydride, and to the formation of the hydride by activation of molecular hydrogen, which is essential for catalysis. The ammonia complex (10) is reported to show some activity towards hydrogenation of maleic, crotonic and acrylic acids [2], but the hydride (10) is not obtained from the Rh(III) ammine and molecular hydrogen, but by zinc reduction of the chloramine salt $[RhCl(NH_3)_5]Cl_2$.

The value of π-acceptor ligands in relation to rapid catalytic activation and transfer of hydrogen was rather clearly established by the observation [3] that chloroplatinic acid in methanol becomes catalytically effective for hydrogenation of ethylene if stannous chloride is added. This observation was extended to include the hydrogenation of acetylene, various terminal alkenes, and styrene [4]. It is likely that the effective species formed is of the type $[PtCl_2(SnCl_3)_2]^{2-}$ and that its catalytic activity is due to the marked π-acceptor capacity of the $SnCl_3^-$ ligands. The stabilising effect of this ligand is also evident when it is noted that tetrachloro*bis* (ethylene) diplatinum (II) is rapidly reduced by hydrogen [5] to form platinum metal:

$$[PtCl_2(C_2H_4)]_2 + H_2 \rightarrow 2Pt + 2C_2H_6 + 4HCl$$

Although $SnCl_3^-$ ligands stabilise the platinum against reduction in this way the chloroplatinic acid-stannous chloride system is not by any means as stable or as useful as phosphine stabilised catalysts, such as $RhCl(PPh_3)_3$.

Semi-quantitatively the influence of π-acceptor capacity of ligands is illustrated in selected systems by the effect of changing the halogen in the group of complexes: $Rh(hal)(PPh_3)_3$, where hal = Cl, Br, I. The rate of hydrogenation of cyclohexene in benzene was found [6] to increase in the order: hal = Cl < Br < I which is also the order of increasing π-acceptor capacity of the halogen. $RhBr(PPh_3)_3$ has been shown [8] to react more rapidly with hydrogen than does $RhCl(PPh_3)_3$, and hydrogenation with $RhBr(PPh_3)_3$ is characterised by smaller enthalpy and entropy of activation terms [7].

In view of the relatively complex set of equilibria which are established in solutions of $Rh(hal)(PPh_3)_3$ under hydrogen (cf. Chapter I) a number of factors may be responsible for the increased rate of reaction with hydrogen and of alkene hydrogenation as the halogen is altered. However, this effect of the halide has some generality since hydride formation by carbonylhalogeno bis (triphenylphosphine) iridium (I) complexes also increases in rate in the order hal = Cl < Br < I, which provides a parallel example [9]:

$$Ir\,hal(CO)(PPh_3)_2 + H_2 \rightarrow IrH_2hal(CO)(PPh_3)_2$$

As indicated in Chapter VII the use of $RhBr(PPh_3)_3$ or $RhI(PPh_3)_3$ may somewhat reduce the chance of alkene isomerisation accompanying hydrogenation [6b], and these derivatives are also rather more soluble than the corresponding chloro complex. There seems, however, to be little general advantage in employing the bromo or iodo complex.

Mention was made [10] in Chapter I of the somewhat greater air stability of the ruthenium carboxylate complexes $RuH(CO_2R)(PPh_3)_3$ in comparison with the chloride $RuHCl(PPh_3)_3$. Variation in the carboxylate group appeared, on the other hand, to have no profound effect on catalytic activity although the trifluoroacetate, $RuH(OCOCF_3)(PPh_3)_3$ seemed least active of those examined [10].

Variation in the basicity of the phosphine ligands in complexes of the type $RhCl(PR_3)_3$ has a considerable effect on the rate of hydrogenation. For aryl phosphines an electron releasing p-substituent increases the rate of alkene hydrogenation, e.g. for $R = p\text{-}XC_6H_4$ the rate increases: $X = H < Me < OMe < NMe_2$. Similarly, electron withdrawing p-substituents, e.g. F or Cl markedly reduce the rate of hydrogenation. Ortho substitution, e.g. by an ortho Me-group retards hydrogenation, no doubt as a result of steric crowding [11].

Alkylphosphines, e.g. PEt_3 are more basic than arylphosphines and by consequently stronger bonding and reduced phosphine dissociation, or by steric crowding, e.g. with $PPh(C_6H_{11})_2$ or $P(C_6H_{11})_3$, alkyl phosphines lead to complexes $RhCl(PR_3)_3$ of low activity [12]. The corresponding phosphite complex, $RhCl[P(OPh)_3]_3$ is also reported not to react with hydrogen [11].

The 5-coordinate *t*-butylphosphine derivative $RhH_2Cl(PBu^t_3)_2$ is stable and despite the bulk of the phosphine ligands is found to be a highly active hydrogenation catalyst [12a]. Also, the complex $RhH(DBP)_4$ which dissociates to $RhH(DBP)_3$ in solution is reported [12b] to be more active in hydrogenation than $RhCl(PPh_3)_3$. The phosphine DBP = 5-phenyl-5H dibenzophosphole, and this instance together with the *t*-butylphosphine complex above also indicates that bulky ligands are not necessarily disadvantageous.

Although, however, alterations in the phosphine ligand may result in appreciable changes in the rate of hydrogenation these are probably concerned with the stage of hydride formation, and the effect on the relative rates of hydrogenation of differently substituted alkenes is small, i.e. changing the phosphine ligand may not very much alter catalyst selectivity [11].

The use of a range of amino phosphines, $PhP(NR_2)_2$ or $Ph_2P(NR_2)$ as ligands in rhodium catalysts $RhClL_2$ or $RhClL_3$ has also been surveyed [12, 13]. Evaluated as catalysts for the hydrogenation of styrene in benzene solution, these complexes show that where the amine component NR_2 is piperidyl, pyrrolidyl, NMe_2 or NEt_2 the derived catalysts have an activity equal to or greater than that of the corresponding triphenylphosphine complex. Where the NR_2 residue is derived from a weak aromatic amine such as PhNHMe or indole the catalyst activity is low. Use of aliphatic amines with bulky alkyl substituents as in ligands such as $PhPN(Pr^i)_2$ or $P(NEt_2)_3$ leads to much reduced catalytic activity. This observation parallels the behaviour of the trialkyl phosphines, noted above, in associating low activity with steric crowding.

1. Solvents

Of the range of solvents which may be used in homogeneous hydrogenation, certain types, which react with the metal hydride are disadvantageous; a wider group of solvents may assist or retard reaction by solvation effects on equilibria, and a small group may make a specific stabilising contribution.

It was shown [6a] that the rhodium complex $RhCl(PPh_3)_3$ may be converted into the derived dihydride by treatment with hydrogen in a range of solvents, e.g. benzene, ethyl acetate, acetic acid, and that the crystalline dihydride $RhH_2Cl(PPh_3)_2$ may conveniently be prepared in chloroform or dichloromethane. However, a chlorinated solvent such as chloroform has been shown [7] to suffer hydrogenation with formation of HCl. This process becomes important under the conditions of hydrogenation catalysis where the concentration of the complex is small and, since HCl is not removed, the catalyst is progressively withdrawn as the relatively inactive hydrogen chloride adduct, $RhHCl_2(PPh_3)_2$.

Inappropriate choice of solvent is found also in the use of alcohols in catalytic deuteriation. It was shown in Chapter VII that with $RhCl(PPh_3)_3$ and deuterium in benzene-methanol or benzene-ethanol the protic solvent is responsible for a good deal of $^1H—^2H$ exchange leading to only partially deuteriated products [14].

The use of alcoholic solvents in combination with $RhCl(PPh_3)_3$ has also been found to lead to a relatively greater amount of alkene isomerisation [15] than when benzene is the solvent (cf. Chapter VII), and also ketal formation in the hydrogenation of keto-containing steroids, particularly in methanol [16] (cf. Chapter III).

The choice of solvent may have an appreciable effect on the rate of hydrogenation. The following data [7] which refer to the hydrogenation of cyclohexene by means of $RhCl(PPh_3)_3$ are illustrative. The rate constant refers to reaction (iii) below.

Solvent	$k \times 10^2$	$\Delta H^{\ddagger}$	$\Delta S^{\ddagger}$
Benzene	31.6 $l\ m^{-}s^{-1}$	18.6 kcal m^{-1}	1.3 e.u.
Benzene-hexane	15.0	23.3	12.9
Hexane-butan-2-one	29.6	16.6	−5.3

Consideration of the detailed reaction sequence for this catalyst shows, however, that the solvent may exert opposing effects at different stages [7]. In assisting phosphine ligand dissociation and solvation of the dihydride a suitable solvent may promote hydrogenation by increasing the available concentration of the reactive hydride $RhH_2Cl(PPh_3)_2(S)$:

$$RhCl(PPh_3)_3 + \text{solvent} \rightleftarrows RhCl(PPh_3)_2(S) + PPh_3 \qquad \text{(i)}$$

$$RhCl(PPh_3)_2(S) + H_2 \rightleftarrows RhH_2Cl(PPh_3)_2(S) \qquad \text{(ii)}$$

On the other hand reaction of the hydride with alkene or other unsaturated acceptor requires solvent displacement:

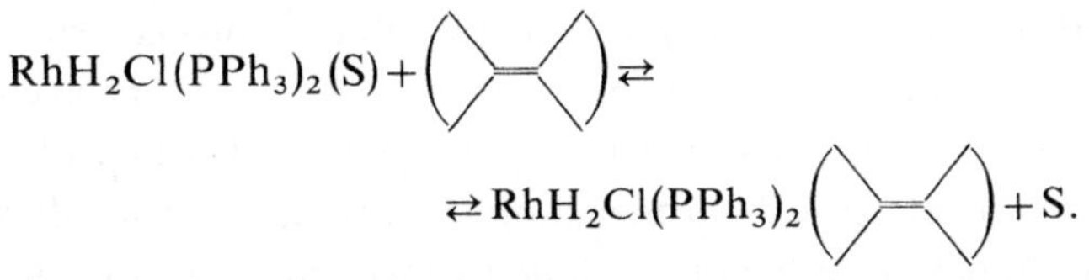

(iii)

In consequence a too strongly co-ordinating solvent may clearly prove inhibitory.

It is also observed [18] that the solvent may influence the relative rates of hydrogenation of e.g. oct-1-ene and hex-1-yne in a mixture. Using $RhCl(PPh_3)_3$ as catalyst in benzene solvent, hex-1-yne and oct-1-ene show a rate ratio in hydrogenation of: hex-1-yne/oct-1-ene = 1.7. This ratio is not materially altered by use of acetone, tetrahydrofuran or dimethylformamide as solvent, but it is increased by use of alcohols, or of polar aprotic solvents such as ethyl acetate or cyclohexanone. The rate ratio is further increased to ca. 5 in chlorobenzene, and to ca. 7 in the rather acidic 2,2,2-trifluoroethanol.

This increased selectivity is associated with increasing rates of hydrogenation of the alkyne, and it seems likely that in different solvents differences in co-ordination power between the alkyne and the alkene become kinetically more significant.

The same study [18] also showed that hydrogenation of oct-1-ene or hex-1-yne in acetic acid, acetonitrile, nitromethane, dimethylformamide or dimethylsulphoxide is very slow. These solvents evidently occupy the solvent co-ordination site in the catalyst to the exclusion of alkene or alkyne, and hence exert the type of retarding effect noted above.

It is clear that with the $RhCl(PPh_3)_3$ catalyst system the solvent has an important function in acting as a displacable ligand. In the case of the group of cationic complexes, $[MH_2(PPh_3)_2S_2]^+$ the solvent S has an equally essential role [19]. Thus the displacement process:

$$[M(diene)\,Cl]_2 + PR_3 \rightarrow [M(diene)\,(PR_3)_2]^+, \qquad M = Rh, Ir$$

is solvent assisted, and particularly by polar solvents such as alcohols or nitromethane, and in the subsequent transformation to the dihydro cationic derivative:

$$[M(diene)\,(PR_3)_2]^+ + H_2 \rightleftarrows [MH_2(PR_3)_2S_2]^+$$

the solvent is co-ordinated. Moreover, n.m.r. examination of salts of $[IrH_2(PPh_3)_2S_2]^+$ shows much more ready solvent exchange where S = acetone than when S is the more strongly co-ordinating acetonitrile. It is significant therefore that whereas the hydride $[RhH_2(PPh_3)_2S_2]^+$, S = Me_2CO, is an effective hydrogenation catalyst in tetrahydrofuran, dioxane, acetone, 2-methoxyethanol or dimethylformamide as solvent, the use of acetonitrile is inhibitory [19]. It is also clear that the carbonyl reducing ability of complexes of this class containing a more basic phosphine than PPh_3 (cf. Chapter VI) is dependent on co-ordination of the ketone through the carbonyl oxygen, i.e. the solvating ligand becomes the reactant [20].

A less well defined solvent influence is evident in the catalytic activity of solutions of $RuCl_3$ reduced in hydrogen in dimethylacetamide [21]. The solvent appears to assist Ru(III)→Ru(II)→Ru(I) reduction, it is thought by acting as a base [21]. A similar interpretation is put on the particular catalytic activity of the complex $[RuCl_2(C_6H_6)]_n$ observed for solutions in dimethylformamide [22], i.e. that the solvent acts as a base in assisting a process:

$$RuCl_2(C_6H_6) + H_2 \rightarrow RuHCl(C_6H_6) + HCl$$

The advantage of using amide solvents was also reported for solutions of $RhCl_3$ in dimethylacetamide which are catalytically active for the hydrogenation of maleic acid and other alkenes [23]. The Rh(III) is reduced by hydrogen to the Rh(I) oxidation level and although co-ordination of the maleic acid certainly contributes to stabilising the rhodium (I) against reduction to the metal, the amide solvent is also important since in aqueous solution $RhCl_3$ is ineffective [24].

As noted in Chapter I, dimethylformamide offers a valuable stabilising solvent medium for catalytically active complexes derived from $RhCl_3py_3$ or $NiCl_2$ reduced *in situ* to a lower oxidation level by means of sodium borohydride [25]. The rhodium catalyst in particular has shown activity for hydrogenation of a wide range of functional groupings. The nature of the stabilising effect of the dimethylformamide is not known, but is is clearly co-ordinated since an active complex $RhCl_2py_2(dmf)(BH_4)$ could be isolated [25]. From a practical point of view, it may also be noted that a considerable proportion of the dimethylformamide solvent may be replaced by ethylene glycol, diethylene glycol dimethyl ether, water or alcohol without causing deposition of rhodium metal during catalysis (Chapter I, Reference 37).

Although the effect of the solvent in homogeneous hydrogenation has not been widely or systematically investigated sufficient is known to point to a rational choice of solvent in particular cases.

2. Ligands and Selectivity

There is some evidence to suggest that substrate selectivity may be usefully influenced by change of ligands. For a series of iridium catalysts, $IrCl(CO)L_2$, for example, it is found that with $L = P(C_6H_{11})_3$, maleic anhydride is hydrogenated much faster than dimethyl maleate, whereas with $L = P(OC_6H_5)_3$ the order of rates is inverted. The effect is even more pronounced with the corresponding iodo complex [26].

References

1. K. Thomas, J. A. Osborn, A. R. Powell, and G. Wilkinson: *J. Chem. Soc. (A)*, 1801 (1968).
2. A. R. Powell: *Platinum Metals Rev.* **11**, 58 (1967).
3. R. D. Cramer, E. L. Jenner, R. V. Lindsey, and U. G. Stolberg: *J. Amer. Chem. Soc.* **85**, 1691 (1963).
4. R. D. Cramer and R. V. Lindsey: *J. Amer. Chem. Soc.* **88**, 3534 (1966); G. C. Bond and M. Hellier: *Chem. and Ind.*, 35 (1965); *J. Catal.* **7**, 217 (1967).
5. J. H. Flynn and H. M. Hulbert: *J. Amer. Chem. Soc.* **76**, 3393 (1954).
6a. J. A. Osborn, F. H. Jardine, J. F. Young, and G. Wilkinson: *J. Chem. Soc. (A)*, 1711 (1966).
6b. G. C. Bond and R. A. Hillyard: *Discuss. Faraday Soc.* **46**, 20 (1968).
7. F. H. Jardine, J. A. Osborn, and G. Wilkinson: *J. Chem. Soc. (A)*, 1574 (1967).
8. W. Strohmeier and R. Endres: *Z. Naturforsch.*, 2613, 362 (1971).
9. L. Vaska and M. F. Werneke: *Trans. New York Acad. Sci.* **31**, 70 (1971).
10. D. Rose, J. D. Gilbert, R. P. Richardson, and G. Wilkinson: *J. Chem. Soc. (A)*, 2610 (1969).
11. C. O'Connor and G. Wilkinson: *Tetrahedron Letters*, 1375 (1969); S. Montelatici, A. van der Ent, J. A. Osborn, and G. Wilkinson: *J. Chem. Soc. (A)*, 1054 (1968).
12. R. Stern, Y. Chevallier, and L. Sajus: *Compt. rend.* **264**, 1740 (1967).
12a. C. Masters, W. S. McDonald, G. Raper, and B. L. Shaw: *J. Chem. Soc. Chem. Comm.*, 210 (1971); cf. C. Masters and B. L. Shaw: *J. Chem. Soc. (A)*, 3679 (1971).
12b. D. E. Budd, D. G. Holah, A. N. Hughes, and B. C. Hui: *Canad. J. Chem.* **52**, 775 (1974); D. G. Holah, I. M. Hoodless, A. N. Hughes, B. C. Hui, and D. Martin, *ibid.*, 3758.
13. Y. Chevallier, R. Stern, and L. Sajus: *Tetrahedron Letts.*, 1197 (1969).
14. W. Voelter and C. Djerassi: *Chem. Ber.* **101**, 58 (1968).
15. R. L. Augustine and J. van Peppen: *J. Chem. Soc. Chem. Comm.*, 495 (1970).
16. W. Voelter and C. Djerassi: *Chem. Ber.* **101**, 1154 (1968).
17. A. S. Hussey and Y. Takencho: *J. Org. Chem.* **35**, 643 (1970).
18. J. P. Candlin and A. R. Oldham: *Discussion Faraday Soc.* **46**, 60 (1968).
19. J. R. Shapley, R. R. Schrock, and J. A. Osborn: *J. Amer. Chem. Soc.* **91**, 2816 (1969); R. R. Schrock and J. A. Osborn: *ibid.* **93**, 2397 (1971).
20. R. R. Schrock and J. A. Osborn: *J. Chem. Soc. Chem. Comm.*, 567 (1970).
21. B. Hui and B. R. James: *J. Chem. Soc. Chem. Comm.*, 198 (1969); *Can. J. Chem.* **52**, 3760 (1974).
22. I. Ogata, R. Owata, and Y. Ikeda: *Tetrahedron Letters*, 3011 (1970).
23. B. R. James and G. L. Rempel: *Discussion Faraday Soc.* **46**, 48 (1968).
24. B. R. James and G. L. Rempel: *Canad. J. Chem.* **44**, 233 (1966).
25. P. Abley, I. Jardine, and F. J. McQuillin: *J. Chem. Soc. (C)*, 840 (1971); P. Abley and F. J. McQuillin: *J. Catal.* **24**, 536 (1972); A. G. Hinze and D. J. Frost, *ibid.*, 540.
26. W. Strohmeier: *Fortschritte der Chemischen Forschung* **25**, 71 (1972).

CHAPTER IX

ASYMMETRIC HYDROGENATION

1. Alkenes

Inspection of the alkene-metal hydride complex (1) in which the ligand X(SML) is asymmetrically substituted, indicates that steric or polar interaction between the groups S, M and L and the alkene substituents a, b, c, d should discriminate between alternative modes of co-ordination of the alkene i.e. between adducts (1) and (2). For this reason the ligand X(SML)

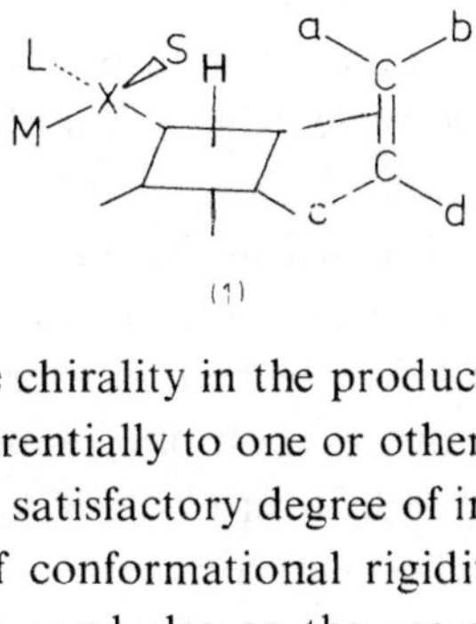

(1)

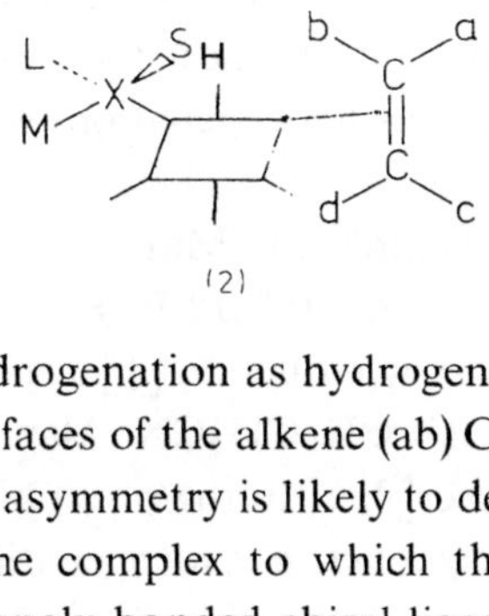

(2)

may induce chirality in the product of hydrogenation as hydrogen is transferred preferentially to one or other of the faces of the alkene (ab) C = C (cd). However, a satisfactory degree of induced asymmetry is likely to depend on a degree of conformational rigidity in the complex to which the alkene co-ordinates, and also on the use of a strongly bonded chiral ligand which does not easily undergo dissociative exchange.

The experimental observation of a useful degree of induced asymmetry in homogeneous hydrogenation has so far been mainly confined to complexes based on rhodium, and has made use of asymmetric ligands of the type (3)–(6). It will be noted that these ligands differ appreciably in the relation of

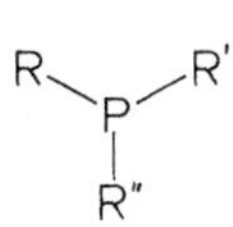

(3)

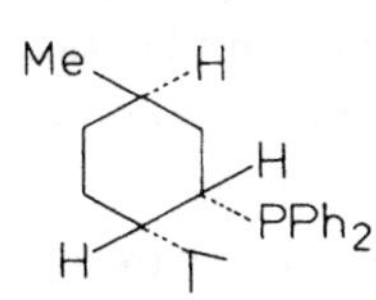

(4)

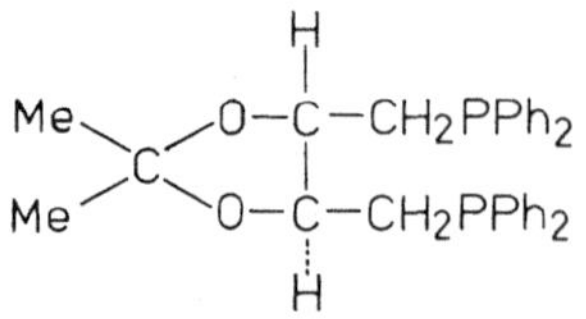

(5)

Ph CH(Me) NH CHO

(6)

the chiral centre to the atom, i.e. phosphorus in (3), (4) and (5), and oxygen in (6), which is co-ordinated to the metal.

Where the asymmetric ligand is a resolved chiral phosphine (3, where R, R′ and R″ are different groups) the centre inducing asymmetry is bonded directly to the metal. The ligands (4), (5) and (6), however, incorporate a chiral carbon centre which is removed by two, three and four atoms respectively from the co-ordinating atom, which in (4) and (5) is phosphorus, and in (6) is the oxygen of the formyl group. Although ligands (4), (5) and (6) are based on readily accessible optically active organic chemical materials, and ligands of type (3) require the synthesis and resolution of asymmetrically substituted phosphines, the first observations of induced asymmetry in homogeneous hydrogenation were in fact made using chiral phosphines of type (3).

A phosphine, e.g. MeEtPhP, may be resolved in the form of the phosphonium salt, $MeEtPhP^+R$, the R substituent being chosen so as to be removable by reduction [1], e.g. $R = PhCH_2$, or by base assisted elimination [2], e.g. $R = CH_2CH_2CN$. Alternatively the resolved phosphine oxide may be reduced [3, 4] to the phosphine using $HSiCl_3$ or $Cl_3SiSiCl_3$.

For asymmetric hydrogenation the rhodium complex is conveniently formed in solution [4] from 2 molar equivalents of the optically active phosphine, e.g. $MePr^nPhP$ and a rhodium complex of the type:

$$[Rh(1,5\text{-hexadiene})\,Cl]_2.$$

Using this system the acid $CH_2{:}C(Ph)\,CO_2H$ was hydrogenated [4] in benzene at 60° and 400 psi H_2 to give $PhCHMeCO_2H$ with some 21% induced asymmetry. A small complication arises from the fact that any excess free phosphine ligand forms a betaine: $MePr^nPhP^+{—}CH_2CHPhCOO^-$ adduct with the unsaturated acid.

Following on these preliminary experiments [5] this procedure has been developed and applied [6] in particular to the hydrogenation of a series of α-acylamino acids (7, R = H or Me, R′ = Ph or Me), or (8, R = Ph or, H, R′ = Ph or Me).

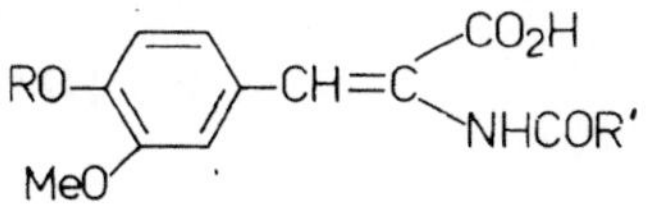

(7)

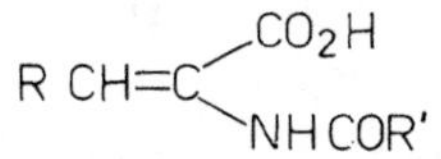

(8)

The degree of asymmetry induced in the hydrogenation products (9) or (10) is modest where the substituent groups R, R′, R″ at the chiral phosphine (11) are e.g. Me, *n*-Pr and Ph, but using larger substituent groups, particularly, R, R′, R″ = cyclohexyl, *o*-anisyl and methyl, leads to products of around 85% optical purity.

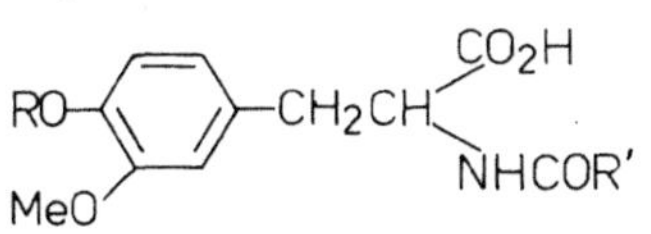

(9)

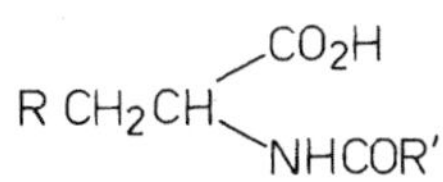

(10)

PRR′R″

(11)

The use of two molar equivalents of the mono-phosphine ligand is clearly necessary, and the geometry of the derived complex has been represented [7] as in (12), i.e. the asymmetry at one ligand centre imposes a preferred orientation on the second.

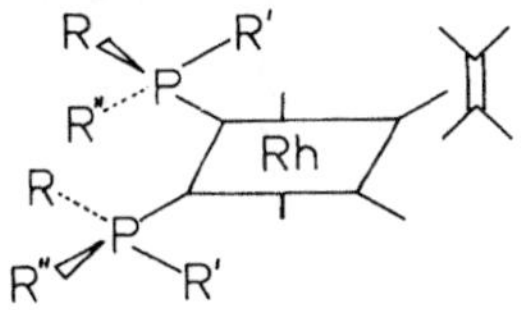

(12)

(e.g. R = Me, R′ = Ph, R″ = *n*-Pr)

There is some support for a formulation (12) since it has been possible to isolate an active complex: $[Rh(\text{cyclo-octa-1,5-diene})L_2]^+BF_4^-$, L = phosphine ligand, carrying two asymmetric phosphines.

A comparable degree of asymmetric induction has, however, also been realised [8] using neomenthyldiphenylphosphine (4) as ligand in combination with $[Rh(C_2H_4)_2Cl]_2$ or $[Rh(\text{diene})Cl]_2$ in benzene-ethanol. Applied to (E)-β-methylcinnamic acid (13), hydrogenation at 60° and 300 psi H_2 for 20 h gave 3-phenylbutanoic acid (14)

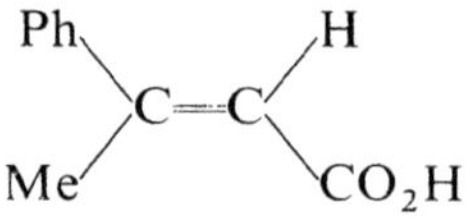

(13)

$PhCHMeCH_2COH$

(14)

containing 61% excess of the (S)-isomer.

This chiral ligand is readily available from (−)-menthol (15) by reaction of $LiPPh_2$ with the corresponding menthyl chloride. It is also to be noted that in the derived rhodium complex the chiral component is the menthyl residue which is further removed from the site of alkene co-ordination than

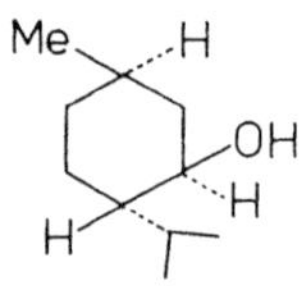

(15)

is the case in complexes where the phosphorus is the centre of asymmetry.

The diphospine ligand (5) (DIOP) represents a particularly useful extension of the concept of using a ligand which is chiral at carbon. The diphosphine is available from the diethyl ester of (L)-(+)-tartaric acid (16) as shown.

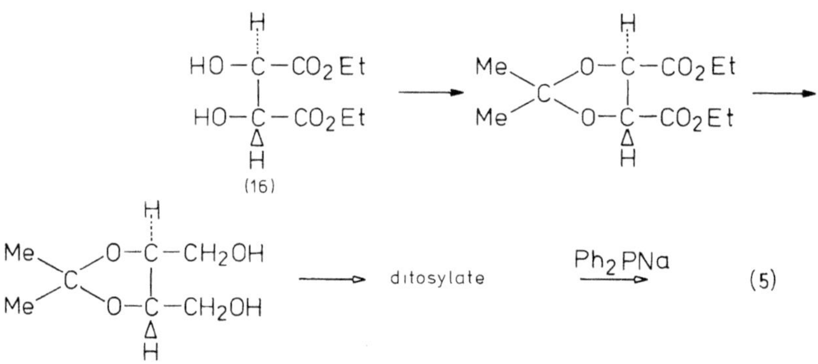

Treatment of $[Rh(cyclo\text{-}octene)_2Cl]_2$ with two molar equivalents of (−)-(5) in benzene gave a solution which proved very efficient for asymmetric hydrogenation[9].

Using this catalyst a series of α-N-acylaminophenylacrylic acid derivatives

(17) could be hydrogenated so as to give the corresponding N-acyl-(R)-phenylalanine (18) R = H, R′ = Me in an optical yield of 70%.

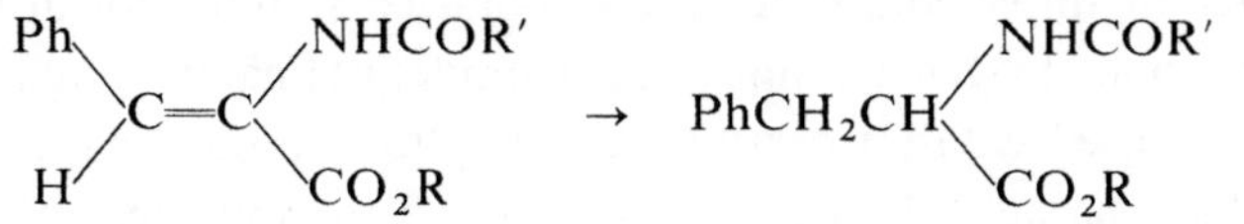

(17) (18)

The optical yield was somewhat smaller for the case where R′ = Ph, R = H, and again for R = Me, R′ = Me. However, the amide (19) was hydrogenated to N-acetyl-(R)-phenylalanamide in 71% optical yield.

Very similar results were also recorded for hydrogenation of the simple acrylic acid derivative (20), and for the substituted aryl derivatives (21),

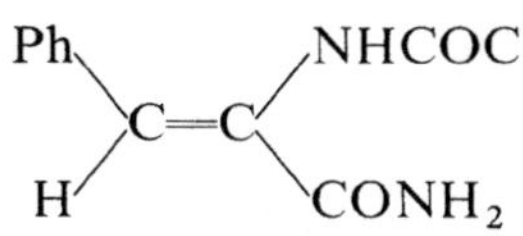

(19)

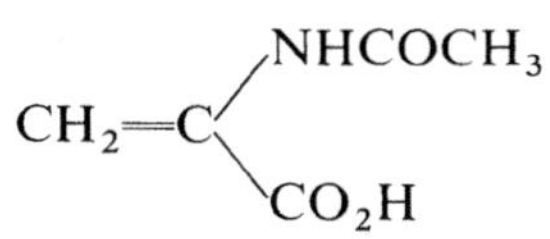

(20)

Ar = *p*-HOC_6H_4 or 3,4-methylenedioxyphenyl. However, (22) was reduced to

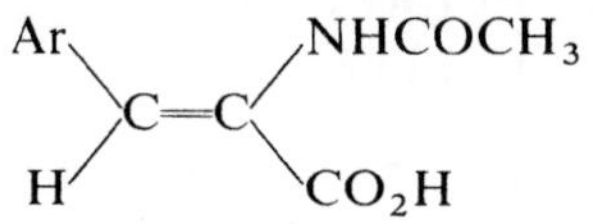

(21)

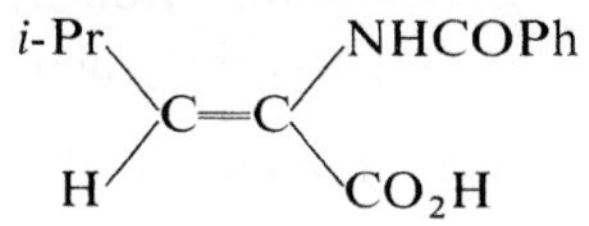

(22)

N-benzoyl-(R)-leucine in an optical yield of only 22%, and (23)–(25) could not be hydrogenated.

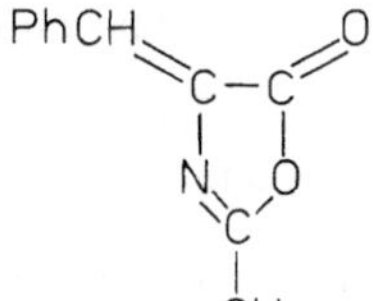

(23)

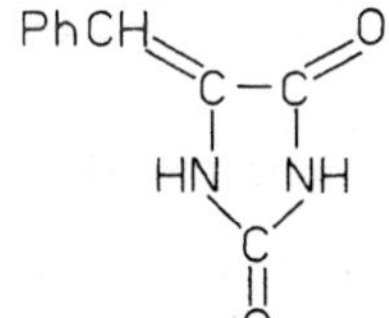

(24)

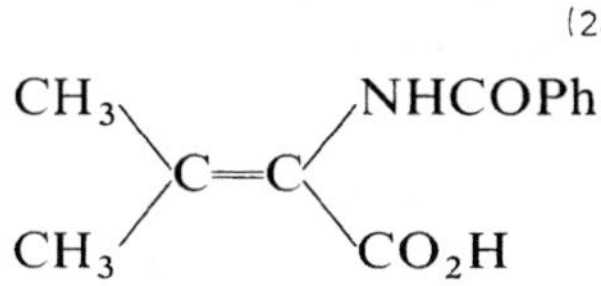

(25)

It has been noted in Chapter VIII that amide solvents such as dimethylformamide or dimethylacetamide provide a medium in which transition metal salts or simple complexes show enhanced stability towards reduction in presence of hydrogen. Complexing of amides as ligands is well known, and a rhodium based catalyst carrying dimethylformamide ligand groups was discussed in Chapter I. Using amides derived from optically active amines as solvent in conjunction with $RhCl_3py_3$ reduced by $NaBH_4$, it has been possible [10] to effect hydrogenation with a useful level of induced asymmetry. Applied to the hydrogenation of methyl 3-phenylbut-2-enoate (26) this method gave the results tabulated below. Using

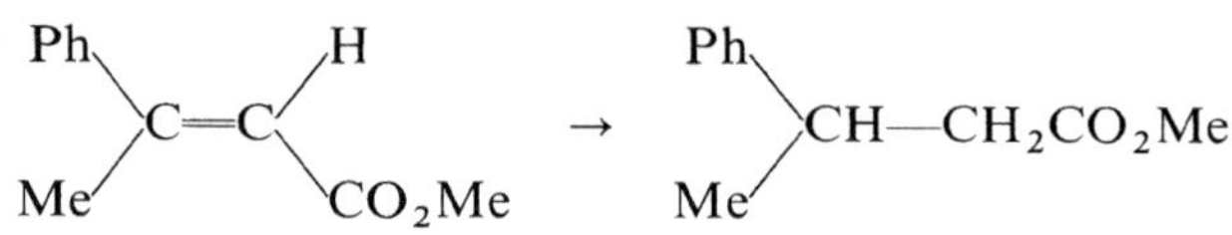

(26) (27)

Amide	*Optical yield*
(+)-PhCHMeNHCHO	(+) 57%
(−)-PhCHMeNHCHO	(−) 48%
(−)-N-formylbornylamine	(+) 48%
(+)-N-acetylglucosamine	(−) 14%

a reasonable structure for the catalyst the configuration of the products could be correlated with the configuration of the chiral centre in the amide [10].

It is interesting that a ligand of this type which must be much less firmly held than a phosphine should prove effective. Also since the amide is presumably co-ordinated through the carbonyl oxygen the chiral centre is rather remote. It is pointed out [10], however, that the 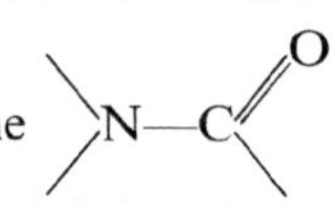 group shows marked rigidity towards rotation about the N—CO bond which may contribute in transmitting asymmetry.

This catalyst system has already found practical application in the asymmetric hydrogenation of folic acid (28) to tetrahydrofolic acid (29) with introduction of asymmetry at * in the product (29). Use of the rhodium catalyst in (−)- or (+)-N-1-phenylethylformamide gave samples of tetrahydrofolic acid of high and low biological activity [11] respectively.

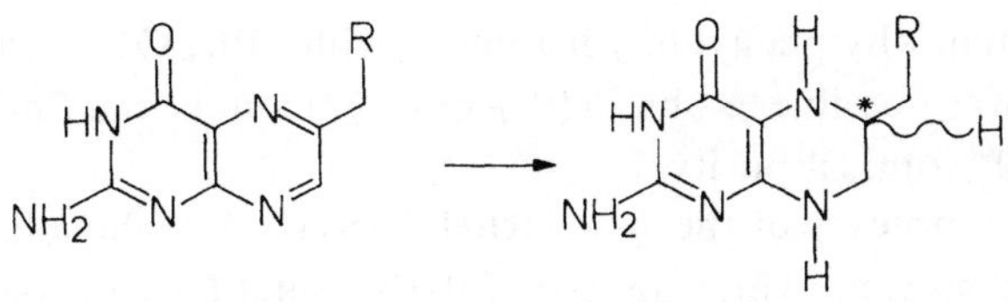

(28) (29)

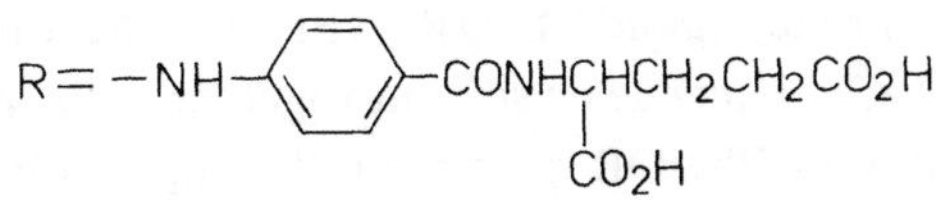

2. Ketones

The hydrosilylation of ketones described in Chapter VI has been adapted to effect asymmetric hydrogenation of the carbonyl group and so to obtain optically active alcohols in useful yield.

The addition of a silane to the carbonyl group is catalysed [12] by $RhCl(PPh_3)_3$ or $RuCl_2(PPh_3)_3$:

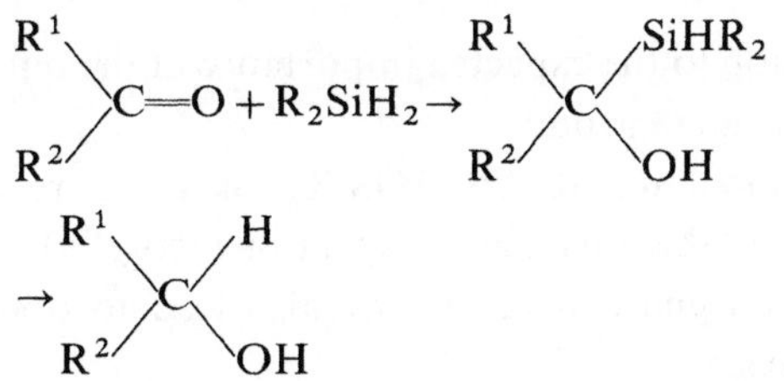

Using $[RhCl(C_8H_{14})_2]_2$ in presence of an asymmetric phosphine, such as (+) or (−)-(30), i.e. (+)- or (−)-DIOP, as catalyst leads [13] to asymmetric

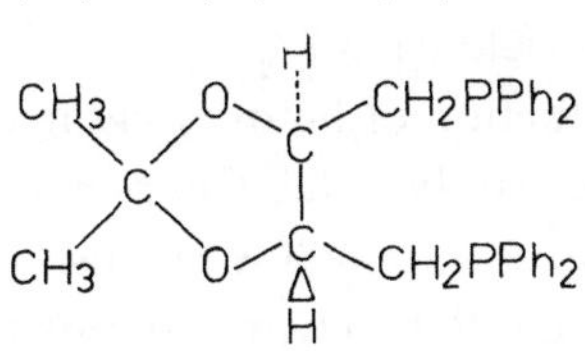

(30)

induction at the hydrosilylation step and thence to optically active alcohols on hydrolysis. It is evident that the chirality is transferred from the asymmetric ligand, such as (30), *via* the silane adduct $R^1R^2C(OH)SiHR_2$, i.e. suitable substitution of the silane will be important.

Using 1-naphthylphenyldihydrosilane with PhCOEt, PhCOMe, or MeCOEt it was possible to obtain (+)- or (−) alcohols, e.g. PhCH(OH) Me, of around 50% optical purity.

Cationic complexes of the type $[RhH_2L_2S_2]^+$, L = phosphine and S = a co-ordinated solvent, which are particularly useful for catalysing hydrogen transfer to a carbonyl group (cf. Chapter 6) have also been applied [14] using a chiral phosphine ligand, L = (R) − $PhCH_2PMePh$. Using this catalyst system in CH_2Cl_2-benzene as solvent together with Me_3SiH or $PhMe_2SiH$, a group of ketones PhCOR gave the following results.

Silane	R	PhCH(OH) R, optical yield
Me_3SiH	Me	5.1%
	Et	6.4
	t-Bu	28.1
	Me	31.6
$PhMe_2SiH$	Et	43.1
	t-Bu	61.8

These data draw attention to the expected importance of the bulk of the substituents both at silicon and carbon.

However, the intermediates $R^1R^2CHOSiR_3$ in these reactions were cleaved using either a suitable Grignard reagent or methyllithium followed by aqueous acid. The Grignard reagent may also lead to optically active silanes in useful yield, e.g.:

$$R'R''CHOSiR^1R^2H \xrightarrow{EtMgBr} EtR'R''SiH$$

(R^1 = 1-naphthyl, R^2 = phenyl) optical yield ca. 30%.

Asymmetric hydrogenation and reduction of ketones *via* hydrosilylation has also been realised using supported catalysts (cf. Chapter XII).

Direct hydrogenation of PhCOMe or $C_2H_5COCH_3$ using as catalyst $[Rh(nbd)\,L_2]^+ClO_4^-$ in ethanol where L is R-benzylmethylphenylphosphine gave 1-phenylethanol enriched in the R-(+) enantiomer and butan-2-ol showing an excess of the R-(−) isomer. However, the optical yields were very small [15]. Somewhat similar results are reported for the hydrogenation of PhCOMe using $[Rh(nbd)(diop)]^+ClO_4^-$ as catalyst [16] in methanol, ethanol or isopropanol. However, this catalyst leads to an appreciably higher degree of asymmetric induction when used[16] for the hydrogenation:

$$\begin{matrix} Ph \\ \\ Me \end{matrix}\!\!\!>\!C{=}N{-}CH_2Ph \rightarrow PhCHMeNHCH_2Ph$$

The asymmetric hydrogenation of imines has also been realised *via* hydrosilylation [17].

References

1. L. Horner, H. Winkler, A. Rapp, A. Mentrup, H. Hoffman, and P. Beek, *Tetrahedron Letters*, 161 (1961).
2. M. J. Gallagher and I. D. Jenkins: *Topics in Stereochemistry* **3**, 1 (1968).
3. L. Horner and W. D. Bulzer: *Tetrahedron Letters*, 1157 (1965).
4. W. S. Knowles, M. J. Sabacky, and B. D. Vinyard: *Ann. New York Acad. Sci.*, 1972, 232 (1970); K. Naumann, G. Zon, and K. Mislow: *J. Amer. Chem. Soc.* **91**, 7012 (1969).
5. W. S. Knowles and M. J. Sabacky: *J. Chem. Soc. Chem. Comm.*, 1445 (1968).
6. W. S. Knowles, M. J. Sabacky, and B. D. Vineyard: *J. Chem. Soc. Chem. Comm.*, 10 (1972); cf. *Adv. Chem. Series* **132**, 274 (1974).
7. L. Horner, H. Siegel, and H. Bushe: *Angew. Chem. Internat. Edn.* **7**, 942 (1968).
8. J. D. Morrison, R. E. Burnett, A. M. Aguiar, C. J. Morrow, and C. Phillips: *J. Amer. Chem. Soc.* **93**, 1301 (1971).
9. H. B. Kagan and T. P. Dang: *J. Amer. Chem. Soc.* **94**, 6429 (1972).
10. P. Abley and F. J. McQuillin: *J. Chem. Soc. (C)*, 844 (1971).
11. P. H. Boyle and M. T. Keating: *J. Chem. Soc. Chem. Comm.*, 375 (1974).
12. R. J. P. Corriu and J. J. E. Moreau: *J. Chem. Soc. Chem. Comm.*, 38 (1973); I. Ojima, M. Nihonyanagi, and Y. Nagai: *ibid.*, 938 (1972); I. Ojima, T. Rogure, M. Nihonyanagi, and Y. Nagai: *Bull. Chem. Soc. Japan* **45**, 3506 (1972); I. Ojima, M. Nihonyanagi, and Y. Nagai: *ibid.*, 3722.
13. R. J. P. Corriu and J. J. E. Moreau: *J. Organomet. Chem.* **64**, C. 51 (1974).
14. K. Yamomoto, T. Hayashi, and M. Kumada: *J. Organomet. Chem.* **54**, C. 45 (1973).
15. P. Bonvicini, A. Levi, G. Modena, and G. Scorrano: *J. Chem. Soc. Chem. Comm.*, 1188 (1972).
16. A. Levi, G. Modena, and G. Scorrano: *J. Chem. Soc. Chem. Comm.*, 6 (1975).
17. N. Langlois, T.-P. Dang, and H. B. Kagan: *Tetrahedron Letters*, 4865 (1973).

CHAPTER X

CATALYTIC ACTIVATION OF ALKANE CARBON-HYDROGEN BONDS

Although as pointed out in Chapter II, there is good evidence for the reversibility of the metal hydride addition step (i) in hydrogenation catalysis, stage (ii) involving hydrogenolysis of the metal alkyl, is effectively non-reversible under these conditions. The reason lies in the thermodynamic stability and lack

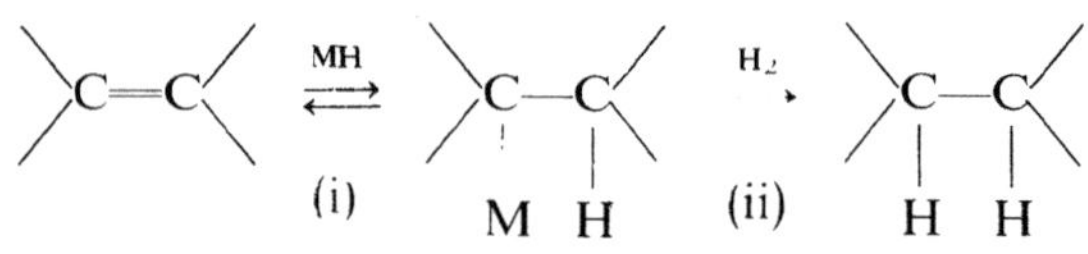

of polarisability of the carbon-hydrogen bond. Despite these effective stabilising factors, however, conditions have in fact been established for activation of the carbon – hydrogen bond for catalytic $^1H—^2H$ exchange in alkanes and cycloalkanes.

The most useful catalytic system employs [1, 2] potassium tetrachloroplatinate (II) in deuterioacetic acid, AcOD, and D_2O in which the alkane is heated for some hours at 100 °C. Under these conditions the platinum salt is unfortunately subject to rather rapid disproportionation: 2Pt(II)→Pt(O)+Pt(IV), and neither Pt(O) nor Pt(IV) is catalytic for $^1H—^2H$ exchange. This deactivating process may, however. be considerably retarded in two ways, by maintaining a high acidity in the medium, conveniently by addition of DCl or better $DClO_4$, and by addition of various aromatic hydrocarbons in small amount [2].

Typically, the alkane or cycloalkane (3 mM) in AcOD and D_2O(1:1, 10 ml) with K_2PtCl_4(0.2 mM) and benzene or pyrene (0.5 mM), and $DClO_4$ (2 mM) is heated for some hours at 100 °C, and for volatile alkanes or cycloalkanes most conveniently in a sealed tube [2]. The presence of air or oxygen is not disadvantageous.

Pyrene is regarded as a more effective stabiliser than benzene [2], but other aromatics have been used [2], and carbazole is also very effective [3].

Under these conditions ^{1}H—^{2}H exchange increases with time [3], e.g. for *n*-decane: after 17 h 2.9, 45 h 9.1, and after 66 h 18.5% ^{2}H incorporation. However, although for smaller alkanes there is evidence of some preference for exchange at CH_3 rather than CH_2 or CH positions, the ^{2}H-*n*-decane showed a distribution between $^{2}H_0$ and $^{2}H_8$ isomers with no obvious selectivity.

Exchange is generally faster for cycloalkanes than for *n*-alkanes of the same number of carbon atoms [2]. In the neighbourhood of chain branching, on the other hand, exchange is largely inhibited. Thus 2,2-dimethylbutane (1) undergoes exchange [2] mainly at position 4, and 2-phenyl-2-methylbutane (2), in the side

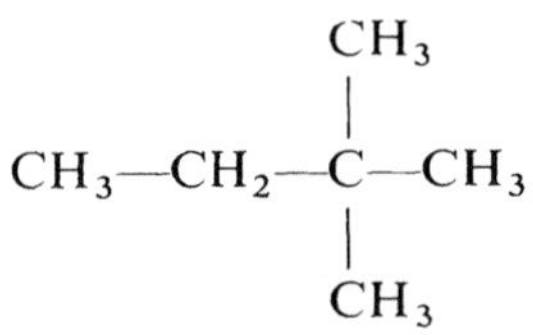
(1)

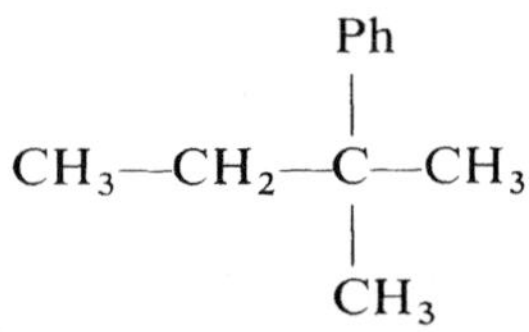
(2)

chain at position 4. Exchange in 3,3-dimethyl-pent-1-ene (3), an olefinic analogue of (2), also occurs at the terminal position 5, as indicated by the ^{1}H n.m.r. spectrum [4]. Similar studies with 3,3-dimethylhex-1-ene (4), however,

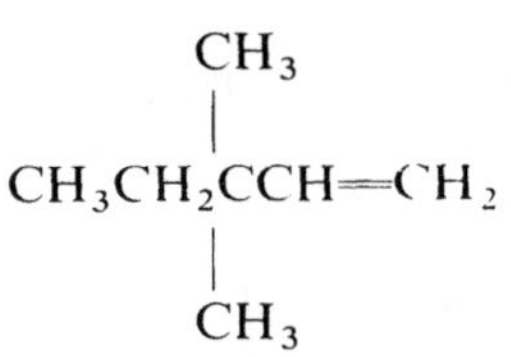
(3)

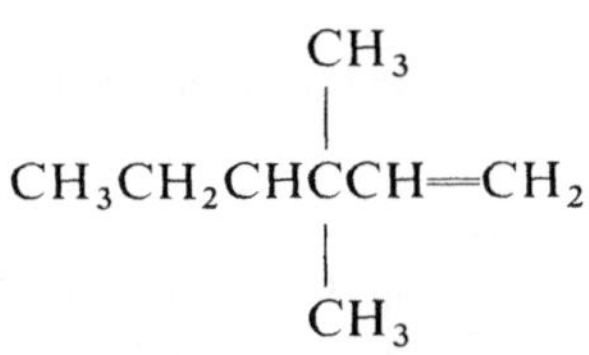
(4)

indicate preferential exchange in position 5, which is non-terminal [14]. This is deduced from the appearance in the proton decoupled ^{13}C n.m.r. spectrum of a 1:1:1 triplet $J_{^{13}C-^{2}H}$ 20 Hz, i.e. indicative of a —CHD— unit [4].

3,3-Dimethylpent-1-ene (3) forms a dichloroplatinum (II) complex (5) which may be isolated [4], and the preference for exchange at position 5 in the chain is ascribed [4] to this co-ordination stabilising exchange *via* a cyclic intermediate (6). Supported for this concept of a cyclic intermediate

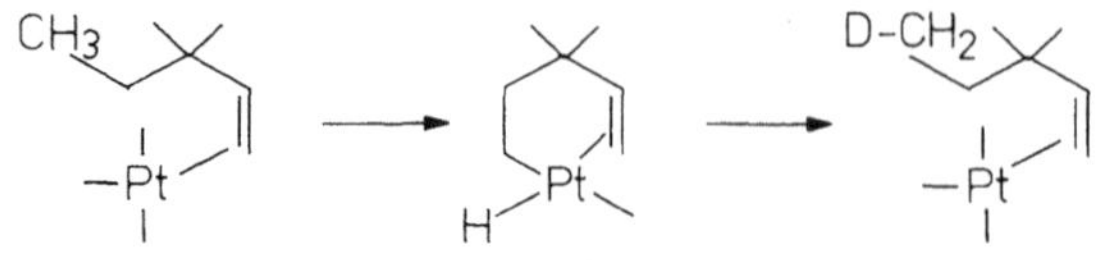

(5) (6) (7)

came [5] from the observation that complexes of the type (8) in $AcOD/D_2O/DClO_4$ at 100 °C, show no ^2H-incorporation in the alkyl group R when $R = C_2H_5$, some 30% ^2H incorporation at the terminal position when $R = CH_2CH_2CH_3$, and 22% ^2H, predominantly (80%) at position 3, where $R = CH_2CH_2CH_2CH_3$. It is suggested that fission of the chlorine bridge to give (9),

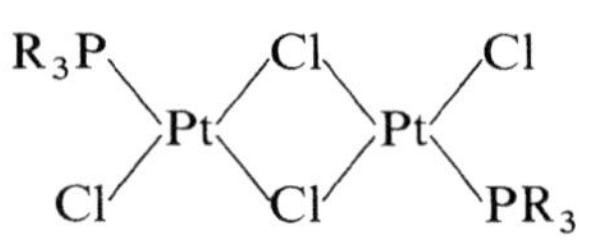

(8)

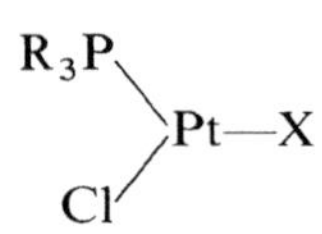

(9)

$X = Cl$ or $ClPtCl_2PR_3$ provides a site for platinum insertion into a CH bond and thence exchange *via* a cyclic intermediate of type (10).

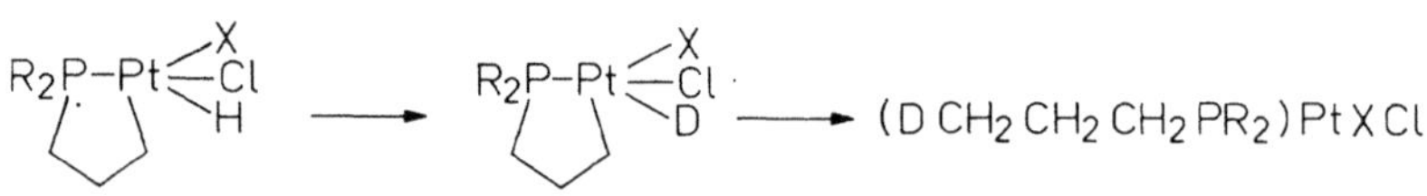

(10)

This type of reaction pattern has many analogies, notably in the well characterised metal insertion into the ortho-CH bond of ligands bearing a phenyl residue [6].

It is to be noted that the $K_2PtCl_4/AcOD/D_2O$ catalyst system also promotes exchange of aryl hydrogen [7], and significantly at rates not very much greater than is found in alkanes or cycloalkanes [2]. However, Na_3IrCl_6, which shows a similar catalytic [8] activity to K_2PtCl_4 in $AcOD/D_2O$, appears to show more discrimination in favour of exchange of alkyl as against aryl hydrogen [8]. Exchange using this iridium catalysed system is, however, considerably slower.

Intermediates such as (6) or (10) are, however, special instances of the type of alkyl platinum intermediate which has been suggested [1, 2] for ^1H—^2H exchange in simple alkanes, which of course do not carry a co-ordinating alkene or phosphine substituent. The reaction is most conveniently expressed as involving an insertion into a carbon-hydrogen bond, followed by exchange of hydrogen on platinum for deuterium, and formation of the ^{2}H-alkane [1, 2] by retransfer of ^{2}H to the alkyl residue:

$$RH + PtCl_2L_n \rightleftarrows RPtHCl_2L_n \xrightleftharpoons{-H^+,\ +D^+}$$
$$RPtDCl_2L_n \rightleftarrows RD + PtCl_2L_n$$
(L = solvent molecules)

Further hydrogen abstraction from the platinum alkyl intermediate to give an alkene-Pt has been suggested [1], but since ^{1}H—^{2}H exchange is observed in cycloalkanes such as adamantane [3], where alkene formation is impossible, this seems unlikely to represent a significant pathway.

For *n*-alkanes from methane to *n*-hexane the rate of ^{1}H—^{2}H exchange increases, and on the basis of an observed linear relation between log rate of exchange and alkane ionisation potential the initial interaction with platinum has been suggested [2] to involve electron transfer from carbon:

$$Pt^{II}Cl_2L_2 + RH \rightleftarrows S_2Cl_2Pt^{I}(\overset{+}{RH}) \rightleftarrows S_2Cl_2Pt^{IV}(H)\ R$$

However, branched chain alkanes do not conform [2] to this rate *vs* ionisation potential relation, and cycloalkanes [2, 3] are also exceptional.

The rate of the exchange reaction is expressed[2]; $d\phi/dt = (k\phi/\phi_\infty)(\phi_\infty - \phi)$ where ϕ is a function expressing the percentage replacement of ^{1}H by ^{2}H, i.e. $\%^2H = \phi/n$ in a hydrocarbon containing n hydrogen atoms, and $\phi = \sum_0^i i\cdot(^2H_i)$. ϕ_∞ is the equilibrium value for exchange. This is important since exchange introduces protons into the medium. However, kinetic examination of the reaction also reveals that the total rate of deuteriation expressed by k_ϕ is greater than the rate of disappearance of $^2H_0$-alkane, defined by the kinetic parameter k_b, viz.:

$$\frac{-db}{dt} = k_b \frac{(b - b_\infty)}{(100 - b_\infty)}$$

when $b = \%\ ^2H_0$-alkane at time t, and $b_\infty = b$, calculated for equilibrium.

This difference, indicating the intervention of an intermediate complex which allows of exchange of more than one hydrogen before the complex dissociates [2], is expressed in terms of a multiple exchange factor $k_\phi/k_b = M$.

For alkanes and simple cycloalkanes M has a value [2] of 1.4–1.9. The phenomenon of multiple exchange may be variously interpreted [2], but what appears to be a significant observation came from a survey [9] of the rates of exchange of cycloalkanes, $CH_2(CH_2)_n$, $n=3$–7, which gave values for the multiple exchange factor:

n	3	4	5	6	7
M	3.6	1.6	1.4	1.4	1.4

Cyclobutane is characterised by carbon-carbon bonds of relatively high p-character. and significantly, cyclopropanes with Pt(II) undergo insertion into a carbon-carbon bond [10]. Also *cis*, anti, *cis*-tetraphenylcyclobutane (11) with K_2PtCl_4 in $AcOH/H_2O$ at 160° undergoes fragmentation to *trans*-stilbene (12). On these grounds the M factor for cyclobutane, and by implication also for other alkanes, is considered [9] to imply a weak bonding asso-

(11) (12)

ciation between Pt(II) and the electrons of the alkane carbon-carbon bonds. Within this association complex a sequence of CH bond insertion, exchange of PtH for PtD, regression to C—C bond interaction etc, may clearly lead to multiple exchange of hydrogen.

Kinetic data [2] indicate that the reagent, initially $PtCl_4^{2-}$, undergoes solvolytic halide displacement to give a catalytically active species $PtCl_2L_2$, where L = solvent, i.e. D_2O, AcOD. Tetrabromoplatinate (II), $PtBr_4^{2-}$ shows only about one tenth of the catalytic activity of $PtCl_4^{2-}$ in the exchange reaction, and the iodoplatinate (II) is inactive [11].

For this reason bromo- and iodo-alkanes fail to show the exchange reaction since rapid halogen interchange between Rhal, hal = Br, I, and $PtCl_4^{2-}$ produces the much less active platinum halide [11]. However, 1-chloro- and 1-fluoro-alkanes undergo exchange, but at rates smaller than for the corresponding alkane, i.e. consistent with the higher ionisation potential [11] due to presence of halogen.

Certain cycloalkanes have been shown to exhibit some regioselectivity in exchange. Adamantane (13) gave a product containing 10.2% ^2H($^2H_0$ 20.0,

$^2H_1$ 30.0, $^2H_2$ 24.9, $^2H_3$ 15.7, $^2H_4$ 6.2, $^2H_5$ 2.4%) with the τ8.1 and 8.2 ^1H n.m.r. signals in a ratio of 1:4.7. This corresponds with preferential exchange of the bridgehead, i.e. tertiary hydrogens [3]. Similarly, norbornane (14) gives a ^2H-norbornane containing 18.9% ^2H, ($^2H_0$ 14.0, $^2H_1$ 20.6, $^2H_2$ 23.8,

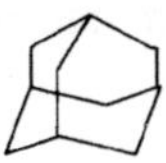

(13)

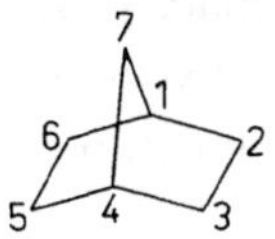

(14)

$^2H_3$ 18.6, $^2H_4$ 13.9, $^2H_5$ 9.1%) showing n.m.r. signals for protons at 1,4(τ 7.8), endo-2,3,5,6 and 7 (τ 8.87) and exo-2,3,5,6 (τ 8.57) in a ratio of 1:3:1 in place of a ratio of 1:3:2 in norbornane [9]. This implies preferential exchange of the exo hydrogens as is also observed [12] in heterogeneously catalysed ^1H—^2H exchange at palladium.

It has been noted [3] that despite the presumed intervention of platinum alkyls, alkane substitution or alkene solvent addition products are noticeably absent in the exchange reaction. Nor does the suggested electron transfer from carbon to platinum appear to lead on to carbenium ions and thence to alkane oxidation products even in cases where the carbenium ion derivative is somewhat stabilised [3]. Exchange is indeed noticeably free of side reactions.

The indicated percentage exchange in the aromatic ring and in the alkyl side chain in toluene (15) shows that aromatic and benzylic hydrogens are of the same order of reactivity in the exchange process [13]. However, com-

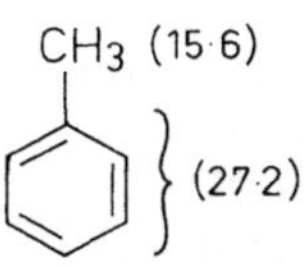

(15)

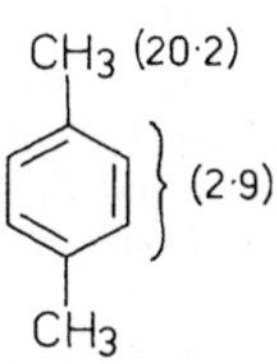

(16)

parable data for *p*-xylene (16) indicate the steric effect of the CH_3-group in hindering exchange in positions ortho to the alkyl substituent [13].

In longer chain alkyl aromatics exchange is observed [13, 14] at all the alkyl CH positions with some preference for the terminal CH_3 as indicated by the %^2H data shown in (17) and (18). However, as noted above, the

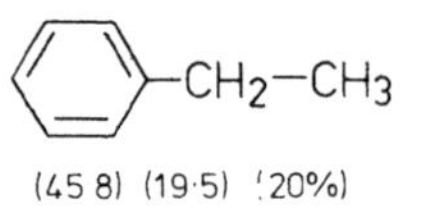

(17)

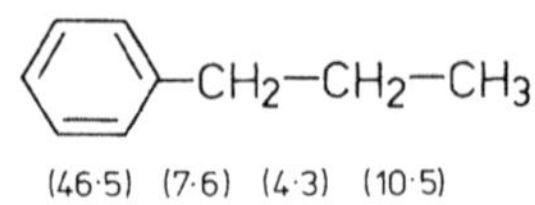

(18)

catalyst system employing Na_3IrCl_6 in place of K_2PtCl_4 may discriminate somewhat in favour of exchange of alkyl hydrogen as indicated at (19) and (20) for % ^2H incorporation at different locations [8].

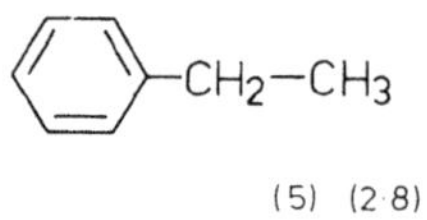

(19)

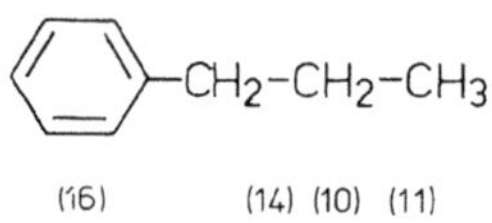

(20)

The kinetic isotopic effect in the exchange process is found [2] to be very similar for alkane (1.7) and arene hydrogens (1.65). However, whilst for an alkane the rate of exchange is reasonably independent of hydrocarbon concentration, or falls slightly as concentration is increased, for the larger aromatics such as naphthalene or pyrene there is a marked inverse relation between rate and hydrocarbon concentration [15].

References

1. N. F. Gol'dshleger, M. B. Tyabin, A. E. Shilov, and A. A. Shteinman: *Russ. J. Phys. Chem.* **43**, 1222 (1969); cf. *Dokl. Akad. Nauk. S.S.S.R.* **198**, 380 (1971).
2. R. J. Hodges, D. W. Webster, and P. B. Wells: *J. Chem. Soc. (A)*, 3230 (1971).
3. G. W. Littlecott and F. J. McQuillin: *Tetrahedron Letters*, 5013 (1973).
4. C. Masters: *J. Chem. Soc. Chem. Comm.*, 1258 (1972).
5. C. Masters: *ibid.*, 191 (1973).
6. Cf. G. W. Parshall: *Acc. Chem. Res.* **3**, 129 (1970); M. I. Bruce: *Organometallic Chem. Chem. Soc.* **2**, 346 (1973); **1**, 282 (1972).
7. Cf. J. L. Garnett: *Catalyst Rev.* **5**, 229 (1971).
8. J. L. Garnett, M. A. Long, A. B. McLaren, and K. B. Peterson: *J. Chem. Soc. Chem. Comm.*, 749 (1973).
9. I. J. Harvie and F. J. McQuillin: *J. Chem. Soc. Chem. Comm.*, 806 (1974).
10. F. J. McQuillin and K. G. Powell: *J. Chem. Soc. Dalton Trans.*, 2123 (1972).
11. R. J. Hodges, D. E. Webster, and P. B. Wells: *J. Chem. Soc. Dalton Trans.*, 2577 (1972).
12. E. McMahon and E. K. A. Clarke: *Tetrahedron Letters*, 1413 (1971).
13. R. J. Hodges and J. L. Garnet: *J. Catalysis* **13**, 83 (1969).
14. J. L. Garnett and R. S. Kenyon: *J. Chem. Soc. Chem. Comm.*, 1227 (1971).
15. R. J. Hodges, D. E. Webster, and P. B. Wells: *J. Chem. Soc. Dalton Trans.*, 2571 (1972).

CHAPTER XI

EXAMPLES OF THE PREPARATION AND USE OF HOMOGENEOUS CATALYSTS

The methods of preparation of some more generally useful homogeneous catalysts are summarised below.

1. Chloro*tris*(Triphenylphosphine)Rhodium(I)[1]

$RhCl(PPh_3)_3$, is obtained by heating rhodium trichloride trihydrate with an excess (ca. 6 mols) of freshly recrystallised triphenylphosphine under reflux in ethanol. The Rh(III) is reduced to the Rh(I) level by the excess of phosphine present which is oxidised to give the phosphine oxide. However, excess phosphine is also important in suppressing the dissociation and dimerisation process:

$$RhCl(PPh_3)_3 \rightleftarrows RhCl(PPh_3)_2 + PPh_3$$
$$RhCl(PPh_3)_2 \rightleftarrows [(Ph_3P)_2RhCl]_2$$

Preparation: To a solution of triphenylphosphine (4 g) in hot ethanol (100 ml) rhodium trichloride trihydrate (0.7 g) in hot ethanol (30 ml) is added in an atmosphere of nitrogen, and the mixture refluxed in nitrogen for 0.5 h. On cooling the product crystallises, and is isolated by filtration and washing with degassed ether, to give red crystals of $RhCl(PPh_3)_3$, (2.0 g), m.p. 159°.

The corresponding bromide may be obtained [1] by halogen exchange without isolating the chloro complex. Thus, to triphenylphosphine (3 g) and rhodium trichloride trihydrate (0.5 g) in hot ethanol (100 ml) under reflux is added lithium bromide (2 g) in hot ethanol (15 ml). After refluxing under nitrogen for ca. 1 h the solution on cooling gives orange crystals of $RhBr(PPh_3)_3$, (1.3 g), m.p. 132° which may be isolated as above.

To prepare the iodo complex $RhI(PPh_3)_3$ the same procedure [1] is used replacing lithium bromide by lithium iodide (2 g) in hot ethanol (40 ml), and the subsequent period of reflux under nitrogen is lengthened to 3–4 h. However, it is necessary in this case to recrystallise the product under nitrogen from the minimum volume of hot benzene containing an excess of triphenylphosphine. Cooling then gives $RhI(PPh_3)_3$ as red crystals, m.p. 118°.

2. Chlorohydrido*tris*(Triphenylphosphine)Ruthenium(II)[2]

$RuHCl(PPh_3)_3$ is obtained by reduction of the corresponding dichloro complex, $RuCl_2(PPh_3)_3$.

Preparation: A solution of ruthenium trichloride (1.0 g) with triphenylphosphine (6.0 g) in ethanol (40 ml) is refluxed under nitrogen for 20 min. Cooling gives brown crystals of $RuCl_2(PPh_3)_3$ which are filtered and washed with a little cold ethanol, m.p. 132°.

Reduction to the hydridochloro complex may be effected by means of hydrogen in the presence of a base, but the more convenient procedure uses sodium borohydride.

Dichloro*tris* (triphenylphosphine) ruthenium (II) (1 g) in hot benzene (125 ml) is treated with a solution of sodium borohydride (0.2 g) in water (2 ml), and the mixture refluxed in a stream of hydrogen or nitrogen. After ca. 1 h the red violet coloured solution is cooled, sodium chloride removed by filtration, and the filtrate evaporated to give $RuHCl(PPh_3)_3$, m.p. 218° in 90% yield.

3. Trichloro*tris*(Pyridine)Rhodium(III)-Sodium Borohydride[3]

This catalyst is prepared *in situ*. Rhodium trichloride trihydrate (1.7 g) in hot ethanol (40 ml) is treated with freshly distilled pyridine (2.4 g) and the solution heated under reflux to give a golden yellow solution. The liquid is decanted from any small amount of solid. Evaporation *in vacuo* and recrystallisation of the residue from the minimum volume of chloroform gives trichloro*tris* (pyridine) rhodium (III) as pale yellow crystals. For hydrogenation the trichloro*tris* (pyridine) rhodium (III) (8–15 mMol l^{-1}) in degassed dimethylformamide (e.g. 30–70 mg in 20 ml dmf) is treated with sodium borohydride (1–2 molar equivalent, e.g. 5 to 10 mg) under hydrogen, to give a brown-pink solution of the catalyst.

4. Nickel Chloride-Sodium Borohydride

This catalyst [4] is prepared by dissolving nickel chloride hexahydrate,

$NiCl_2$, $6H_2O$ (30 mg) in degassed dimethylformamide (20 ml) and adding sodium borohydride (4 mg) under hydrogen.

5. Carbonylhydrido*tris*(Triphenylphosphine)Rhodium(I)[5]

Rhodium trichloride is reduced by triphenylphosphine in hot ethanol in the presence of formaldehyde which by decarbonylation acts as a source of carbon monoxide.

A boiling solution of triphenylphosphine (3.6 g) in ethanol (150 ml) is slowly treated with a solution of rhodium trichloride trihydrate (1 g) in ethanol (35 ml), and, after the solution has cleared, sufficient 37% aqueous formaldehyde (5–10 ml) is added to change the colour of the solution from red to pale yellow. The product obtained on cooling is filtered, washed with a little ethanol and then with diethylether and dried *in vacuo*. Recrystallisation from the minimal volume of hot benzene gives *trans*-carbonylchloro*bis* (triphenylphosphine) rhodium (I) (2.2 g), m.p. 195–197°. It is not, however, necessary to isolate this chloro-complex which may be reduced directly by addition of sodium borohydride. To the hot yellow solution prepared above is added sodium borohydride (1 g) in ethanol (20 ml). Filtration gives carbonylhydrido*tris* (triphenylphosphine) rhodium (I) which is washed with a little ethanol and dried *in vacuo*, yield 2.2 g.

6. Cationic Complexes $[MH_2L_2S_2]^+$

Complexes of this type, M = Rh, Ir, L = PR_3, AsR_3, S = solvent [6, 7] are obtained by treating a diene complex: $[M(diene)L_2]^+$ with hydrogen. A number of such derivatives have been prepared [6]. In a typical example diene = norbornadiene and L = triphenylphosphine.

Chloro*tris* (triphenylphosphine) rhodium (I) (11.0 g) with norbornadiene (1.5 ml) and potassium hexafluorophosphate (3.1 g) in acetone (50 ml) and dichloromethane (10 ml) was stirred for 2 h. The reaction mixture is then filtered and the potassium chloride on the filter washed free of orange material by means of dichloromethane. Evaporation of the filtrate to crystallisation followed by addition of an equal volume of ethanol and excess diethyl ether gave $[Rh(NBD)(PPh_3)_2]^+PF_6^-$, 9.5 g.

In an alternative procedure [7] $[Rh(NBD)Cl]_2$ is treated with $AgPF_6$ in acetone under an inert atmosphere (argon) to give $[Rh(NBD)]^+PF_6^-$ which is then treated with the phosphine.

Corresponding iridium complexes $[Ir(diene)L_2]^+PF_6^-$ have also been made in a similar manner. However, the rhodium derivatives appear to be the more effective in hydrogenation catalysis.

Cationic complexes of the type $[Rh(COD)_2]BF_4$ also give solvent co-ordinated derivatives such as $[Rh(COD)(MeCN)_2]^+$, COD = cyclo-octa-1,5-diene, which are effective catalysts for hydrogenation [8]. Preparation starts from e.g. Rh(COD)(acac) which (2.6 g) in methylene dichloride (20 ml) is treated with cyclo-octa-1,5-diene (5 ml) and then triphenylmethyl tetrafluoroborate (3.0 g) in methylene dichloride (40 ml) dropwise. Addition of diethyl ether to the red solution gives *bis* (cyclo-octa-1,5-diene) rhodium tetrafluoroborate (3.2 g) m.p. 206°. This material (1.4 g) in dichloromethane (20 ml) treated dropwise with acetonitrile (5 ml) gives a yellow solution from which ether (100 ml) precipitates *bis* (acetonitrile) (cyclo-octa-1,5-diene) rhodium (I) tetrafluoroborate (1.15 g), m.p. 188°.

7. Hydridopentacyanocobaltate (III)

The hydridopentacyanocobaltate (III) ion has been isolated [9] as a mixed caesium sodium salt, $Cs_2Na[Co(H)(CN)_5]$, by reducing a solution of cobalt (II) chloride (0.5 M) and sodium cyanide (2.5 M) with sodium amalgam (0.5% Na), adding 2 vols, air free alcohol to precipitate impurities, followed by caesium chloride solution (3 M) to give the hydrido complex as a solid. This material shows Co—H bands at 1860 and 774 cm^{-1}.

Alternatively [10] a lithium salt $Li_3CoH(CN)_5$ may be isolated as a solvate by shaking a solution of $Li_3Co(CN)_5$ (0.5 M) in alcohol at 0° under hydrogen for 10 hours, and precipitating with a mixture of acetone and ether (2:1). The product which is very moisture sensitive is dried in nitrogen.

For hydrogenation it has been customary [11] to use the complex as $K_3Co(CN)_5$, in aqueous solution, e.g. by mixing equal volumes of aqueous cobalt (II) chloride (0.3 M) and aqueous potassium cyanide (1.54 M) solutions, under hydrogen.

8. Tricarbonylchromium-Methyl Benzoate [12]

Hexacarbonylchromium (10 g) and methylbenzoate (60 ml) are heated under reflux in nitrogen for 12 h. After cooling, excess methyl benzoate is removed in vacuo (25°/0.5 mm Hg), and the residue sublimed at 80–90° and 0.1 mm Hg to give tricarbonylchromium-methyl benzoate (8.5 g) as an orange red solid, m.p. 93°.

The experimental conditions used in a number of examples of homogeneous hydrogenation are described below.

9. Using Chloro*tris*(Triphenylphosphine)Rhodium (I)

(i) *Nootkatone (1)→11,12-Dihydro nootkatone (2)* (Reference 13)

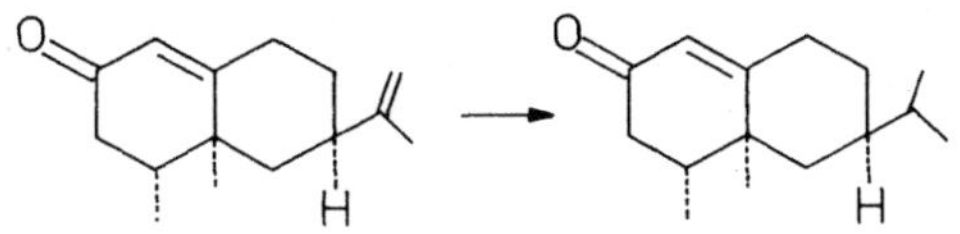

(1) (2)

Nootkatone (1.0 g) in dry benzene (100 ml) with $RhCl(PPh_3)_3$ (0.8 g) shaken under hydrogen took up 1 molar equivalent in 1 h. The solution was percolated through alumina (40 g), and which was then eluted with benzene (4 × 25 ml). Evaporation of the benzene solution and distillation of the residue gave (2), 0.95 g, b.p. 120°/0.025 mm Hg.

(ii) (±)-*Longicamphenilone (4)* (Reference 14)

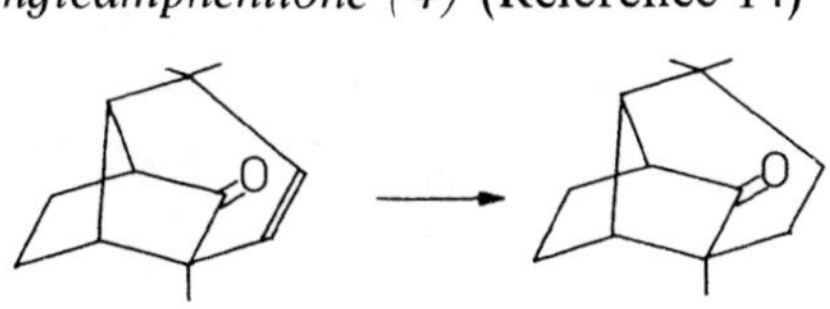

(3) (4)

$RhCl(PPh_3)_3$, (75 mg) in benzene-methanol (3:1, 50 ml) was shaken under hydrogen for 45 min and the alkene (3) (115 mg) in benzene-methanol (3:1, 40 ml) injected *via* a septum. After stirring under hydrogen for 2.5 days the solution was percolated through basic alumina and the eluate evaporated to give (±)-longicamphenilone (4), (100 mg).

(iii) *Dehydrogriseofulvin (5)→griseofulvin (6)* (Reference 15)

Dehydrogriseofulvin (0.5 g) in benzene (20 ml) with $RhCl(PPh_3)_3$ (0.1 g) was shaken under hydrogen at 55–60° for 2 days. Chromatography and crystallisation of the product gave griseofulvin (0.2 g).

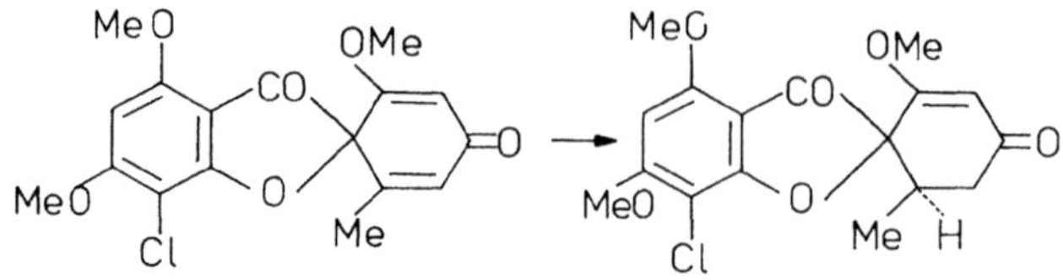

(5) (6)

(iv) *7-Isopropenyl-10-methyltricyclo ([4, 4, 0, 0^{1,5}])-decan-4-one (7)→ norcubebanone (8)* (Reference 16)

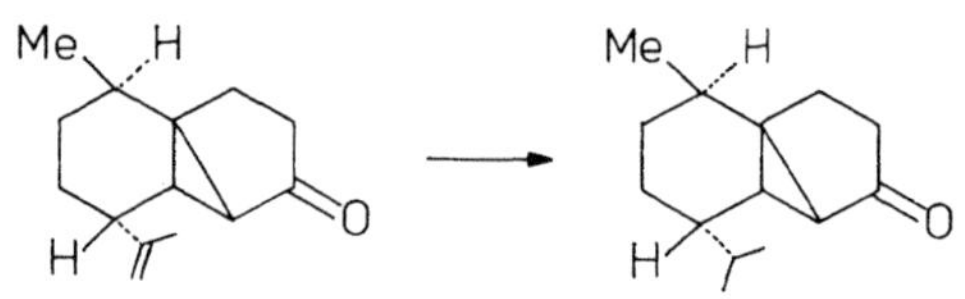

(7) (8)

The ketone (7) (70 mg) with $RhCl(PPh_3)_3$ (50 mg) in benzene (10 ml) was hydrogenated and the solution then passed through a short column of neutral alumina. The eluate afforded (8) (68 mg) which crystallised.

(v) *Thebaine (9)→8,14-dihydrothebaine (10),* (Reference 17)

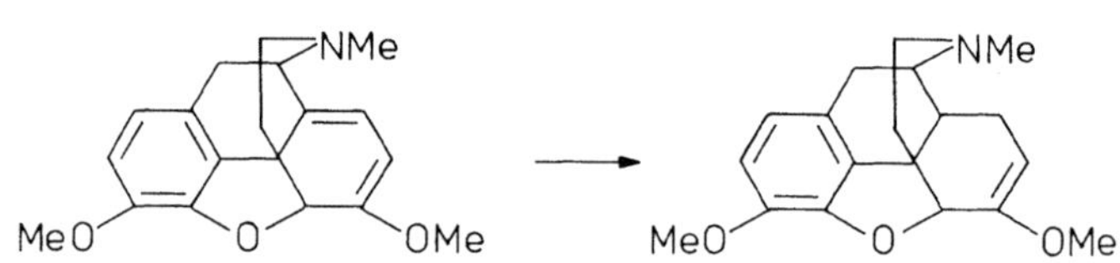

(9) (10)

Thebaine (1 g) with $RhCl(PPh_3)_3$ (0.15 g) in benzene (30 ml) took up 1.12 mols of hydrogen in 20 h. Chromatography on alumina and crystallisation gave 8,14-dihydrothebaine (10) (0.77 g), m.p. 164°.

(vi) *Steroid-1,4-diene-3-ones (11)→steroid-4-en-3-ones (12)* (Reference 18)

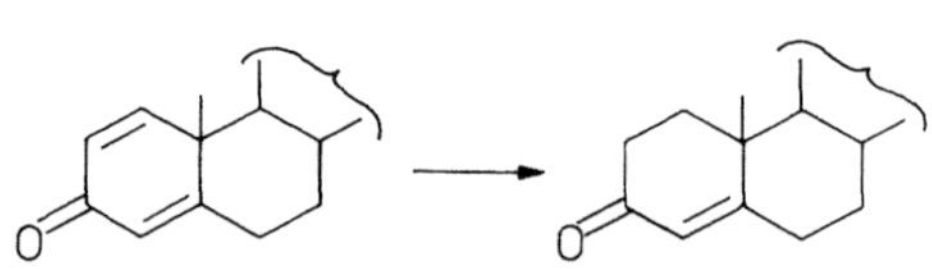

(11) (12)

The steroid-1,4-diene-3-one (11) (0.1 g) was dissolved in a 5mM solution of $RhCl(PPh_3)_3$ in benzene (20–50 ml) and shaken under hydrogen for some hours [14].

(vii) *α-Santonin (13)→1,2-dihydro-α-santonin (14)* (Reference 19)

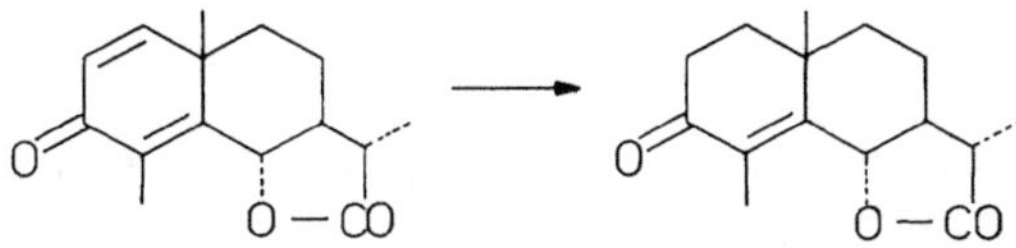

(13) (14)

Dihydrosantonin (14) required in the course of a synthesis of (−)-frulanolide and (+)-arbusculin B was prepared as follows.

α-Santonin (5 g) in benzene and ethanol (1:1, 250 ml) with $RhCl(PPh_3)_3$ (100 mg) was stirred under hydrogen for 12 h. The theoretical volume of hydrogen had then been absorbed. After removal of solvent the residue was chromatographed on neutral alumina (50 g). Elution with ethylacetate-benzene (1:9) gave an eluate which was recrystallised from hexane-carbon tetrachloride to give dihydro-α-santonin (14), 4.95 g, m.p. 102°.

10. Using Dichloro*tris*(Triphenylphosphine)Ruthenium (II)

Androsta-1,4-diene-3,17-dione (15)→androst-4-en-3,17-dione (16) (Reference 20).

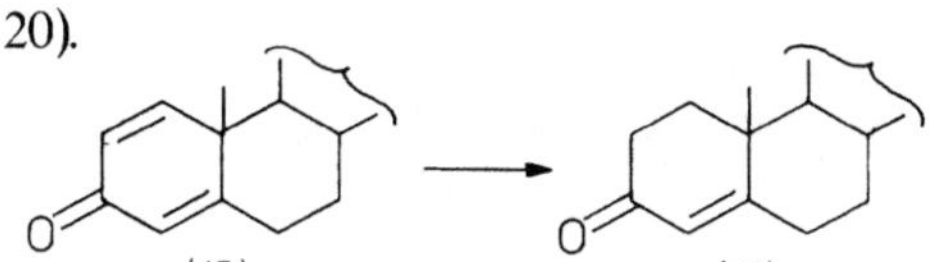

(15) (16)

The yield of (16) in comparison with the yield of androstan-3,17-dione, which is formed by further reduction, is found to increase with hydrogen pressure when hydrogenation is carried out in presence of $RuCl_2(PPh_3)_3$. Good results were obtained using androsta-1,4-dien-3,17-dione (0.75 g) with $RuCl_2(PPh_3)_3$ (0.15 g) in benzene (15 ml) at 40° under 125 kg cm^{-2} H_2 for 10 h. This gave a product yielding (16) in 94% yield.

11. Using the Trichloro*tris*(Pyridine)Rhodium (III)–Sodium Borohydride Catalyst

Steroid-4-en-3-ones (17)→5αH- and 5βH-steroid-3-ones (18), (19)

(Reference 21)

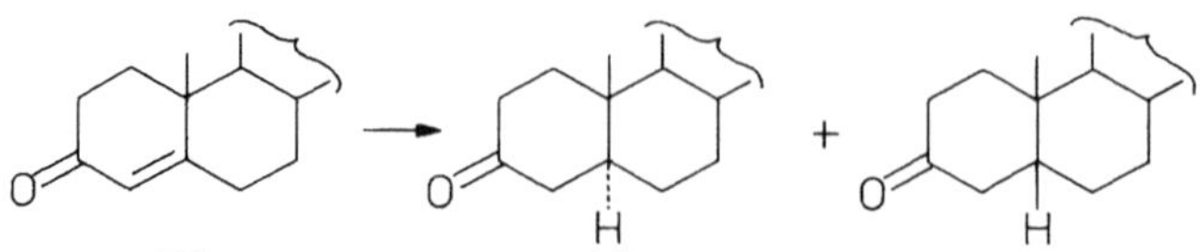

(17) (18) (19)

The steroid-4-en-3-one at a concentration of 2×10^{-2} M in dimethylformamide containing $RhCl_3py_3$ and $NaBH_4$ at 7.5×10^{-3} M concentration was shaken under hydrogen. The theoretical uptake of hydrogen was completed in 6–10 h. at the ordinary temperature. The products were separated by t.l.c. on silica gel. The steroids hydrogenated in this way were cholest-4-en-3-one, testosterone, progesterone, and 17-methyl testosterone.

12. Using Nickel Dichloride-Sodium Borohydride (Reference 22)

To a stirred solution of $NiCl_2$ (0.4 mMole) in dimethylformamide (100 ml) at 25° is added sodium borohydride (0.44 mMole) and when the solution has become brown hydrogen is admitted followed by the alkene (25 mMole).

13. Using Pentacyanocobaltate (II) (Reference 11)

(20) (21) (22)

A solution of $K_3Co(CN)_5$ (200 ml, 0.15 M) was shaken under hydrogen to equilibrium, 260 ml H_2 being absorbed. Norbornadiene (0.02 mole) (20) was then injected in portions. Isolation of the hydrocarbon product by means of ether gave material containing norbornene (21) and nortricyclene (22) in a ratio of 1:3 as indicated by g.l.c. analysis.

14. Using Tricarbonylchromium-Methyl Benzoate (Reference 23)

Hydrogenations are carried out in a 150 ml capacity magnetically stirred autoclave [24] with the substance to be hydrogenated (10 mMol) in *n*-hexane (50 ml) and tricarbonylchromium-methyl benzoate (0.5 mMol). The autoclave is flushed with nitrogen and then repeatedly with hydrogen before heating in hydrogen (30 atm.) to 160°.

15. General

A number of these hydrogenation catalysts are air sensitive, particularly in solution, and require care in use if the optimal activity is to be realised. Of those listed $RhCl(PPh_3)_3$, $RuCl_2(PPh_3)_3$ and the cationic complexes $[ML_2S_2]^+$ are particularly oxygen sensitive. Solutions of the pentacyanocobaltate (II) ion are subject to 'aging' and must be freshly prepared.

References

1. J. A. Osborn, F. H. Jardine, J. F. Young, and G. Wilkinson: *J. Chem. Soc. (A)*, 1711 (1966); J. A. Osborn and G. Wilkinson: *Inorg. Synth.* **10**, 67 (1967).
2. P. S. Hallman, B. R. McGarvey, and G. Wilkinson: *J. Chem. Soc. (A)*, 3143 (1968); P. S. Hallman, J. A. Stephenson, and G. Wilkinson: *Inorg. Synth.* **12**, 237 (1970); R. A. Schunn and E. R. Wonchoba: *ibid.* **13**, 131 (1972).
3. P. Abley, I. Jardine, and F. J. McQuillin: *J. Chem. Soc. (C)*, 840 (1971).
4. P. Abley and F. J. McQuillin: *J. Catalysis* **29**, 536 (1972); A. G. Hinze and D. J. Frost: *ibid.*, 540.
5. D. Evans, G. Yagupsky, and G. Wilkinson: *J. Chem. Soc. (A)*, 2660 (1968); cf. D. Evans, J. A. Osborn, and G. Wilkinson: *Inorg. Synth.* **11**, 99 (1968); N. Ahmad, S. D. Robinson, and M. F. Uttley: *J. Chem. Soc. Dalton Trans.*, 843 (1972).
6. J. R. Shapley, R. R. Schrock, and J. A. Osborn: *J. Amer. Chem. Soc.* **91**, 2816 (1969); R. R. Schrock and J. A. Osborn: *ibid.* **93**, 2397 (1971); *idem., J. Chem. Soc. Chem. Comm.*, 567 (1970).
7. R. R. Schrock and J. A. Osborn: *J. Amer. Chem. Soc.* **93**, 3089 (1971).
8. M. Green, T. A. Kuc, and S. H. Taylor: *J. Chem. Soc. Chem. Comm.*, 1553 (1970); *idem., J. Chem. Soc. (A)*, 2334 (1971); M. Green and T. A. Kuc: *J. Chem. Soc. Dalton Trans.*, 832 (1972).
9. R. G. S. Banks and J. M. Pratt: *J. Chem. Soc. (A)*, 54 (1968).
10. G. Pregaglia, D. Morelli, F. Conti, G. Gregorio, and R. Ugo: *Discuss. Faraday Soc.* **46**, 110 (1968).
11. J. Kwiatek and J. K. Seyler: *Advances in Chem. Series*, No. 70, 207 (1968).
12. R. B. King: in J. J. Eisch and R. B. King (eds.), *Organometallic Syntheses*, Vol. 1, Acad. Press, New York, 1965, p. 136.
13. H. C. Odom and A. R. Pinder: *J. Chem. Soc. Perkin I Trans.*, 2193 (1972).
14. J. E. McMurry and S. J. Isser: *J. Amer. Chem. Soc.* **94**, 7132 (1972).
15. A. J. Birch and K. A. M. Walker: *Australian J. Chem.* **24**, 513 (1971).
16. A. Tanaka, R. Tanaka, H. Uda, and A. Yoshikoshi: *J. Chem. Soc. Perkin I. Trans.*, 1721 (1972).
17. A. J. Birch and K. A. M. Walker: *J. Chem. Soc. (C)*, 1894 (1966).

18. C. Djerassi and J. Gutzwiller: *J. Amer. Chem. Soc.* **88**, 4537 (1966).
19. A. E. Greene, J.-C. Muller, and G. Ourisson: *J. Org. Chem.* **39**, 186 (1974).
20. S. Nishimura and K. Tsuneda: *Bull. Chem. Soc. Japan* **42**, 852 (1969).
21. P. Abley, I. Jardine, and F. J. McQuillin: *J. Chem. Soc. (C)*, 840 (1971).
22. A. G. Hinze and D. J. Frost: *J. Catalysis* **24**, 540 (1972).
23. E. N. Frankel and R. O. Butterfield: *J. Org. Chem.* **34**, 3930 (1969).
24. E. N. Frankel, E. A. Emken, and V. L. Davison: *J. Org. Chem.* **30**, 2739 (1965).

CHAPTER XII

SUPPORTED CATALYSTS

As molecular complexes of definite structure, catalytically active metal derivatives are in principle more easily reproducible in their properties than a heterogeous metal catalyst. Moreover, by variation in the metal of the complex and the co-ordinating ligands some systematic variation is possible in their reactivity and selectivity as catalysts. Despite these merits, however, a soluble catalyst complex may present problems in relation to recovery of the hydrogenation products and of the catalyst itself. Reference to Chapter XI where examples of the use of homogeneous catalysts are described shows that in laboratory work the product and soluble catalyst have frequently been separated by chromatography, whereas similar reactions effected using a heterogeneous catalyst call only for separation by filtration.

For greater convenience in separation of the reaction products the catalytically active complex may be bonded to an insoluble solid support. This device is also worth study on other grounds. Suitably supported organometallic complexes may prove more effective, as for example the benzylzirconium derivatives bound to silica in alkene polymerisation [1]. Moreover, study of supported complexes where the surface structure can be examined or deduced may advance knowledge in relation to hetereogeneous catalysis.

In the studies so far reported two main types of catalyst support have been investigated, organic polymers such as polystyrene or polyvinyl chloride, or silica. The chemical problem is to provide specifically reactive sites on the polymer or silica surface to which the metal complex may be co-ordinated.

The procedure adopted with polystyrenes has been to chloromethylate [2] the aryl residues in the polystyrene and treat the product with a diarylphosphine salt such as $(C_6H_5)_2PLi$. The resulting polymer bearing diphenylphosphine substituents then co-ordinate to a suitable transition metal which is presented in solution.

$$]\text{-Ar} \xrightarrow[\text{SnCl}_4]{\text{MeOCH}_2\text{Cl}}]\text{-ArCH}_2\text{Cl} \xrightarrow{\text{LiPPh}_2}]\text{-ArCH}_2\text{PPh}_2$$

This procedure has been applied [3] to bind rhodium from a solution of $RhCl_3$ to give a polymer bound rhodium phosphine complex which was effective for hydrogenation of for example hept-1-ene to heptane, crotonaldehyde to butyraldehyde, and of other substances at 65° and 24 atm H_2. However, there is little information regarding the nature of the surface complex or of the oxidation level of the rhodium. A better defined process allows the polystyrene-CH_2PPh_2 material to exchange with the chloro*tris* (triphenylphosphine) rhodium (I) complex in solution. However, this ligand exchange may be slow and far from complete due to accumulation of triphenylphosphine in solution. The resultant catalyst polystyrene-$[CH_2PPh_2RhCl(PPh_3)_2]$ was, however, shown to be effective for the hydrogenation of a range of simple olefins [4].

In another approach [5] a polystyrene was brominated in the aromatic rings (bromine, ferric bromide), the halogen replaced by lithium using *n*-butyllithium in tetrahydrofuran, and this lithio derivative then treated with diphenylchlorophosphine. The polymer-diphenylphosphine, treated with $M(CO)Cl(PPh_3)_2$, M = Rh or Ir, or with $RhCl(C_2H_4)(PPh_3)_2$, yielded polystyrene supported rhodium or iridium complexes.

An alternative procedure [6] was directed to phosphenylation of styrene before polymerisation using a reaction sequence:

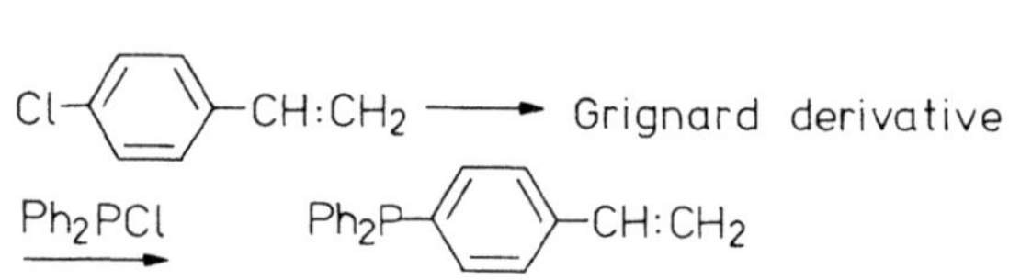

Dispersion polymerisation of this product in presence of divinylbenzene gave a phosphenylated polymer which with $[(C_2H_4)_2RhCl]_2$ in dimethylformamide yielded a rhodium bearing polymer with evolution of ethylene. This preparation was shown to be active for hydrogenation catalysis.

Polymethacrylic esters in which the ester group includes a suitable ligand atom have also been employed [7] to bind platinum or rhodium, e.g. polymers of (1) or (2):

$$CH_2=CMeCOOCH_2CH_2Me_2 \tag{1}$$

$$CH_2=CMeCOOCH_2CH_2CH_2CH_2PPh_2 \tag{2}$$

In a more systematic study [8] other workers have treated a diphenylphosphine derived from a chloromethylated polystyrene or from a vinyl chloride polymer with $Rh(CO)_2(acac)$, $[RhCl(COD)]_2$, $Ni(COD)_2$, $NiCl_2$ and $CoCl_2$. These supported metal complexes were shown to be active catalysts for e.g. hydrogenation or hydroformylation.

The same authors [8] have also explored an interesting method for binding a phosphine to the surface of silica. This makes use of 2-diphenylphosphinoethyltriethoxysilane which with tetracarbonyldichlorodirhodium (I) yields a rhodium complex (3):

$$(EtO)_3SiCH_2CH_2PPh_2+[RhCl(CO)_2]_2$$
$$\rightarrow RhCl(CO)[Ph_2PCH_2CH_2Si(OEt)_3]_2 \tag{3}$$

By the action of sodium borohydride in ethanol (3) gave rise to the derived hydrido complex (4).

$$RhH(CO)[Ph_2PCH_2CH_2Si(OEt)_3]_2 \tag{4}$$

On refluxing (3) or (4) in toluene with silica the surface hydroxyl groups displace one or more of the ethoxy residues on silicon thus binding the metal complex (3) or (4) to the surface *via* —SiOSiO— bonds. Alternatively the $Ph_2PCH_2CH_2Si(OEt)_3$ reagent may first be allowed to react with the silica, and the resultant phosphine bearing solid used in the ligand displacement reaction with a metal complex.

A polymer supported titanium catalyst which is effective for hydrogenation of e.g. hex-1-ene, cyclohexene, hex-3-yne, styrene or diphenyl acetylene at 1 atm hydrogen pressure and the ordinary temperature has also been reported [9].

The procedure for preparing this catalyst started from a chloromethylated polystyrene which by the action of sodium cyclopentadienyl gave the derivative (5). Treatment of this cyclopentadienyl with methyllithium followed by $(cp)TiCl_3$ gave the complex (6) which was activated by treatment with butyllithium.

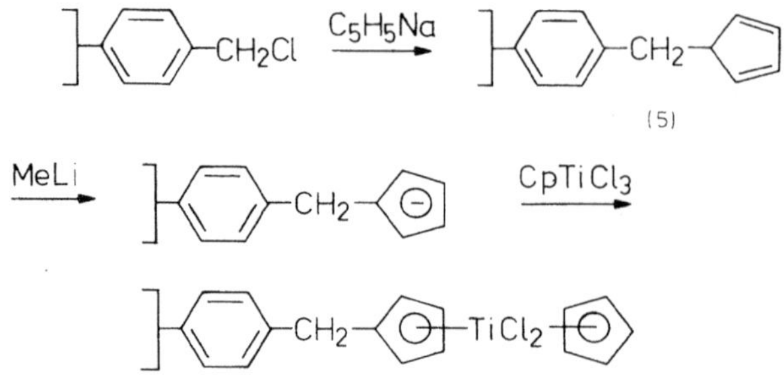

(5)

(6)

The methods of asymmetric induction applied to alkene hydrogenation or hydrosilylation of ketones described in Chapter IX have also been extended to the use of supported catalysts [10].

A chloromethylated styrene-divinylbenzene co-polymer (a Merrifield resin [11]) as in (7) may be oxidised as shown to the aldehyde derivative (8). This resin derivative is then suitable for combination with the chiral diol (9) *via* cyclic acetal formation (10). The diol (9) is derived as indicated, from

hydrolysed.

(7) (8)

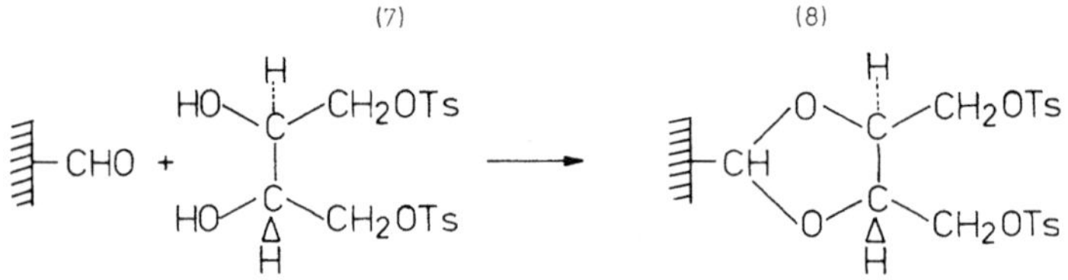

(9) (10)

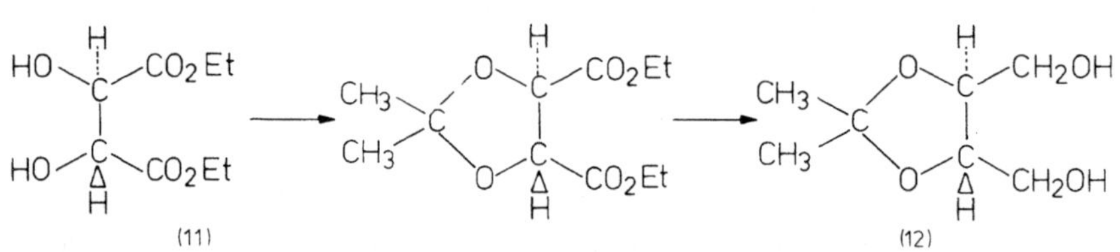

(11) (12)

The di-toluene-*p*-sulphonate (10) with Ph_2PLi gave the *bis*-phosphine containing resin (13), which was found to abstract rhodium from a benzene solution of $[RhCl(C_2H_4)_2]_2$ to give a resin supported catalyst (14).

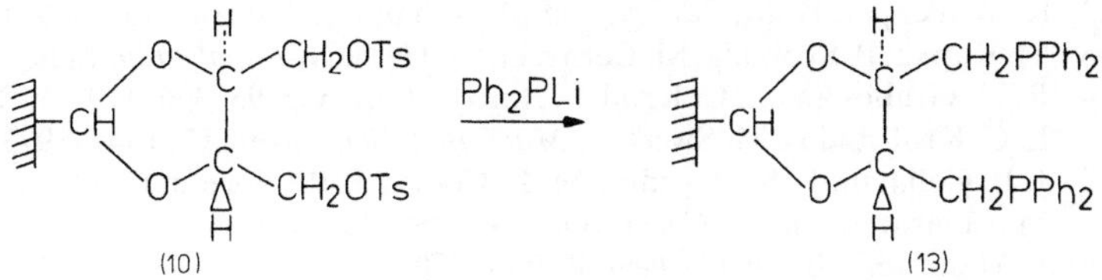

(10) (13)

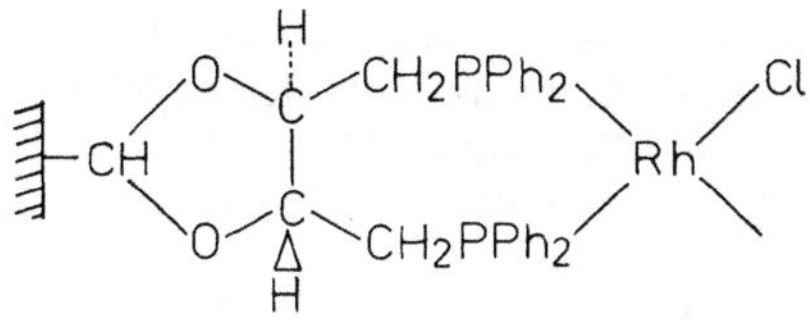

(14)

This product proved to be an efficient hydrogenation catalyst, but showing a remarkable difference in the degree of asymmetry induced in alkene hydrogenation and ketone hydrosilylation.

In alkene hydrogenation, e.g. with PhC(Et):CH_2 or PhC(CO_2Et):CH_2 the optical activity induced in the PhCH(Et) CH_3 or PhCH(CO_2Et) CH_3 product was very small and the acid: PhCH:C(CO_2H) NHCOMe failed to hydrogenate. However, using the catalyst for hydrosilyation of e.g. PhCOMe gave, after acid hydrolysis, (S)-(−)-PhCH(OH) Me in ~50% optical purity using a catalyst based on the phosphine derived from (+)-tartaric acid.

The evident advantage of the procedure is that the valuable asymmetric catalyst may be recovered by filtration and reused.

Preparation of a catalyst in a molten salt medium has also been explored as a means of establishing a phase boundary [12]. However, the experimental conditions are compatible only with relatively rather stable non polar substances. Various alkenes have been hydrogenated using e.g. $PtCl_2$ in a medium of fused $(C_2H_5)_4N^+$ $SnCl_3^-$ at 140° and 100 atm pressure of hydrogen. The alkenes studied include cyclododecatriene which under optimum conditions may be hydrogenated reasonably selectively to cyclododecene (cf. however Chapter XIII). The hydrocarbon may be separated from the salt medium as a second phase or by distillation.

References

1. Cf. D. G. H. Ballard: *Chemistry in Britain* **10**, 20 (1974).

2. K. W. Pepper, H. M. Paisley, and M. A. Young: *J. Chem. Soc.* 4097 (1953).
3. M. Kapka, P. Svoboda, M. Cerny, and J. Hetflejs: *Tetrahedron Letters*, 4787 (1971).
4. R. H. Grubbs and L. C. Kroll: *J. Amer. Chem. Soc.* **93**, 3062 (1971); R. H. Grubbs, L. C. Kroll and E. M. Sweet: *J. Macromol. Sci. Chem.* **A7**, 1047 (1973).
5. J. P. Collman, L. S. Hegedus, M. P. Cooke, J. R. Norton, G. Dolcetti, and D. N. Marquardt: *J. Amer. Chem. Soc.* **94**, 1789 (1972).
6. J. Manassen: *Israel J. Chem.* **8**, 5p (1970).
7. M. Kapka, P. Svoboda, M. Kraus, and J. Hetflejs: *Chem. and Ind.*, 650 (1972).
8. K. G. Allum, R. D. Hancock, S. McKenzie, and R. C. Pitkethly: *Proceedings of the 5th International Congress on Catalysis* (1972).
9. R. H. Grubbs, C. Gibbons, L. C. Kroll, W. D. Bonds, and C. H. Brubaker: *J. Amer. Chem. Soc.* **95**, 2373 (1973).
10. W. Dumont, J.-C. Poulin, T.-P. Dang, and H. B. Kagan: *J. Amer. Chem. Soc.* **95**, 8295 (1973).
11. R. B. Merrifield: *J. Amer. Chem. Soc.* **85**, 2149 (1963); **86**, 304 (1964).
12. G. W. Parshall: *J. Amer. Chem. Soc.* **94**, 8716 (1972).

CHAPTER XIII

HOMOGENEOUS VERSUS HETEROGENEOUS HYDROGENATION

The long established heterogeneous metallic hydrogenation catalysts have the advantage of being effective with a wide range of groupings capable of accepting hydrogen, and they offer the convenience that the catalyst is easily removed from the reaction mixture by filtration. This wide utility, however, is an indication of some lack of selectivity, and, in addition, metallic catalysts, notably palladium, may induce alkene bond migration and isotope scrambling in deuteriation, and unwanted hydrogenolysis.

The available homogeneous catalysts are to a considerable extent complementary to their heterogeneous counterparts. They are characterised by marked selectivity in the groups hydrogenated, and by a limited tendency to isomerise alkenes. Hence deuteriation is commonly well controlled and stereospecific. Further the homogeneous catalysts which have been examined show very limited activity in hydrogenolysis.

There is, however, the practical disadvantage (cf. Chapter XI), that recovery of the hydrogenation products may commonly require chromatography, and the catalytic complex may consequently be effectively lost. It is likely that this disadvantage may in time be overcome by the developments discussed in Chapter XII.

An instance which illustrates the selectivity of the homogeneous type of catalyst is provided by the results of hydrogenation of α-santonin (1). α-Santonin with platinum or palladium [1] is hydrogenated non-selectively to mixtures of isomeric tetrahydrosantonins (2). α-Santonin may be hydro-

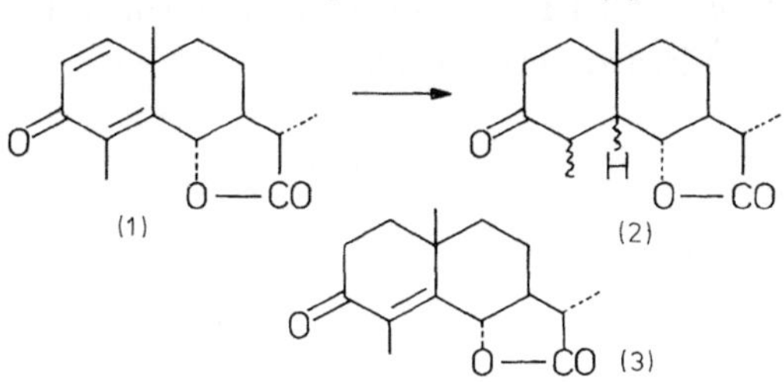
(1) (2) (3)

genated selectively using Raney nickel in benzene [2] to give dihydrosantonin (3), but some workers [3] found the yield by this procedure low. Using the homogeneous catalyst $RhCl(PPh_3)_3$ for hydrogenation, however, dihydrosantonin (3) becomes readily available in good yield [4].

A further example which illustrates the control of hydrogenation achieved by the use of $RhCl(PPh_3)_3$ as catalyst is provided by dehydrogriseofulvin (4). At variously supported palladium, or ruthenium catalysts, or at platinum [5], dehydrogriseofulvin is hydrogenated with extensive hydrogenolysis to the benzophenone derivative (5). However, using $RhCl(PPh_3)_3$, dehydrogriseofulvin is hydrogenated [6] in good yield to griseofulvin (6).

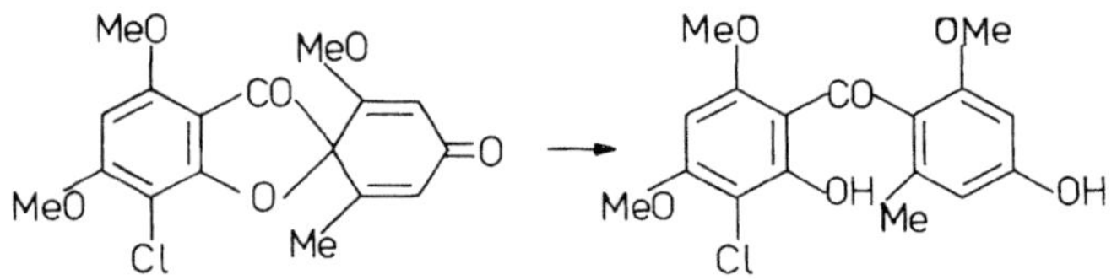

(4) (5)

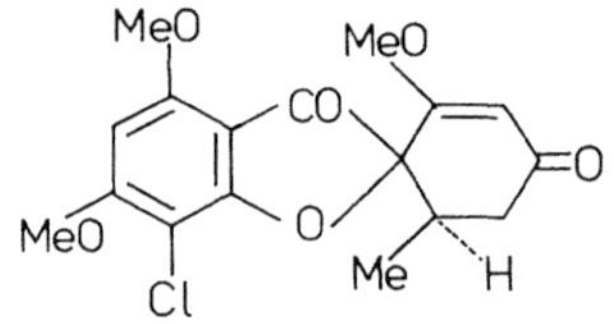

(6)

It appears from this example that the heterogeneous metallic catalysts may co-ordinate strongly to the oxygen function which initiates hydrogenolysis, whilst $RhClH_2(PPh_3)_2$ preferentially transfers hydrogen to the alkene centre.

A formally similar example is offered by eremophilone (7) which at palladium on various supports and in a variety of solvents is hydrogenated preferentially at the carbonyl conjugated alkene bond [7]. This preference is, however, reversed using $RhCl(PPh_3)_3$ as catalyst [7] which therefore offers a selective route to the dihydro-derivative (8).

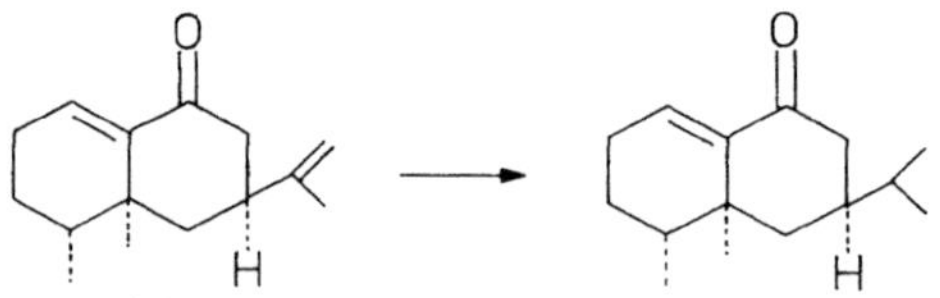

(7) (8)

This observation provided the precedent for the use of $RhCl(PPh_3)_3$ as catalyst to effect [8] hydrogenation of nootkatone (9) to the dihydro-derivative (10).

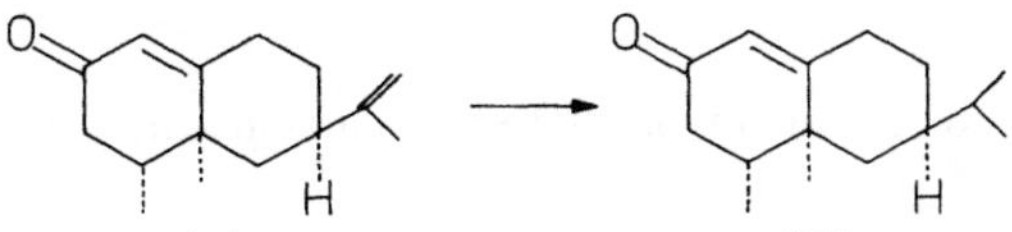

(9) (10)

Discrimination between an alkene and a conjugated enone grouping:

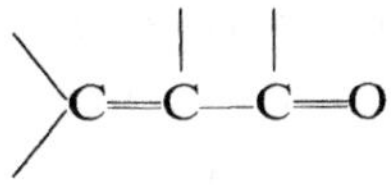

may also be achieved for heterogeneous hydrogenation, but in the opposite sense to that observed in examples (7) and (9). It is part of the empirical art of heterogeneous hydrogenation that the rate of hydrogenation of an alkene bond is much reduced in an alkaline medium, e.g. ethanolic potassium hydroxide, whereas the rate of hydrogenation of an enone moiety is very much less affected [9], Using this device cyperone (11) for example may be hydrogenated [10] to the tetrahydro-derivatives (12) at palladium/charcoal in ethanol and selectively at the enone grouping with palladium on charcoal in 2% ethanolic potassium hydroxide to give (13).

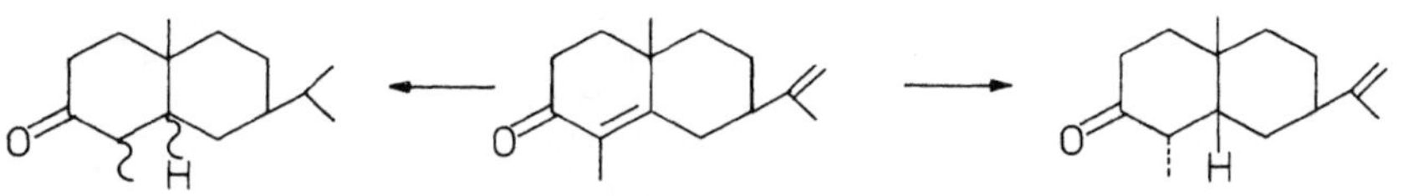

(12) (11) (13)

Thus heterogeneous and homogeneous hydrogenation may offer complementary techniques in realising selective hydrogenation.

An impressive instance of selectivity which may be achieved using a homogeneous catalyst is provided by the hydrogenation of the cyclododeca-1,5,9-trienes (14). The desired result is to effect hydrogenation to cyclododecene (15) which is a starting material for manufacture of polyamides based on a C_{12}-unit. From a thorough survey [11] it appears that this result can be achieved using various ruthenium complexes e.g. $Ru(CO)_3(PPh_3)_2$ in benzene which gave 94.6% of (15), or $RuCl_3(Et_2S)_3$ in dimethylformamide yielding 92.6% of (15). It may also be noted that the solvent used may be very important in determining selectivity of hydrogenation to give cyclododecene.

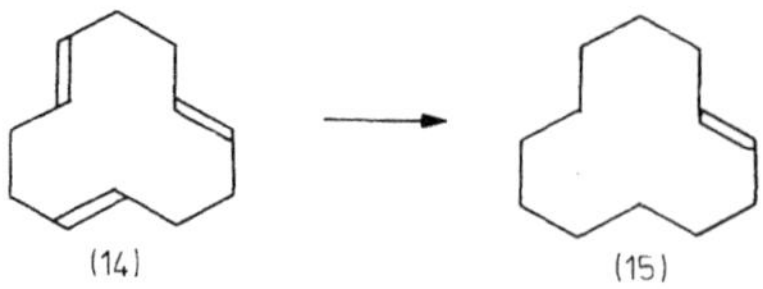

(14) (15)

Heterogeneous hydrogenation of cyclododecatriene fails to exhibit this type of selectivity [12].

A particular advantage of homogeneous catalysis is found in relation to controlled deuterium addition to an alkene (cf. Chapter VII). Although the resistance of heavily substituted alkenes to homogeneous hydrogen or deuterium addition is a limitation, homogeneously catalysed deuterium addition is generally much more specific than is the case with corresponding heterogeneously catalysed reactions.

This may be illustrated by reference to the case of cholesteryl acetate (16) which with deuterium in AcOD and platinum gave [13] a $^2H_2$-cholestanyl acetate (17) containing 2.55 moles ^2H/mole [13]. The series of transformation indicated below located the additional deuterium at the 7-position, i.e. platinum catalyses exchange of the allylic hydrogens.

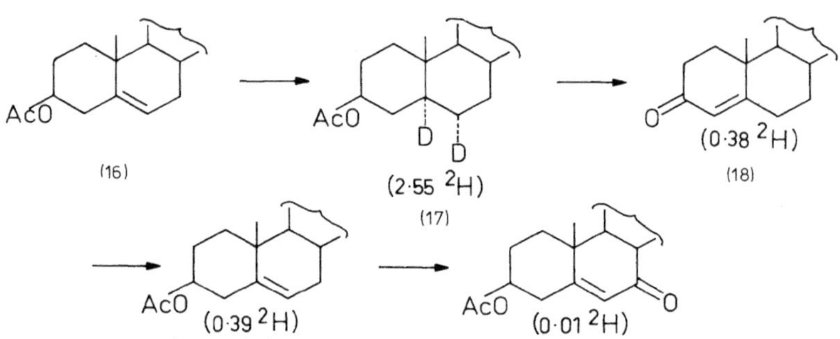

(16) (17) (18) (19) (20)

An alkene bond located as in (16) is not hydrogenated by homogeneous catalysts such as $RhCl(PPh_3)_3$, but it is interesting to note that 17β-hydroxyestra-4,6-dien-3-one (21) in dioxan with this catalyst and deuterium gave [14] stereospecifically a product (22) showing a ^2H content: $^2H_0$ 3, $^2H_1$ 4, and $^2H_2$ 93%.

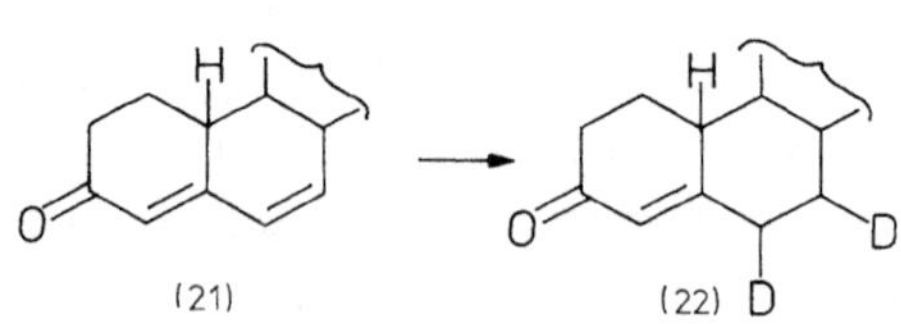

(21) (22)

It is also reported [15] that there is a difference between the stereochemistry of deuterium addition to androst-1,4-dien-3.17-dione (23) catalysed by $RhCl(PPh_3)_3$ which gives the $1\alpha,2\alpha$-$^2H_2$ adduct in good yield, and by palladium which leads to the $1\beta,2\beta$-$^2H_2$ product in low yield.

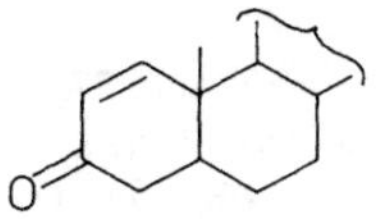

(23)

The range of activity of the more useful homogeneous catalysts may be summarised.

Catalyst	Groups hydrogenated
$RhCl(PPh_3)_3$*	$RCH{=}CH_2 >$ *cis* $RCH{=}CHR >$ *trans* $RCH{=}CHR > R_2C{=}CHR$, allenes, carbonyl *via* hydrosilylation
$RuHCl(PPh_3)_3$	$RCH{=}CH_2 \gg RCH{=}CHR$
$RuH(OCOR)(PPH_3)_3$	$RCH{=}CH_2 \gg RCH{=}CHR$
$RhH(CO)(PPh_3)_3$	$RCH{=}CH_2 \gg RCH{=}CHR$
$RhCl_3py_3/NaBH_4$	Alkenes, Ar-carbonyl, Ar- or allyl-halides, epoxides, RNO_2, heteroaromatics.
η^3-$C_3H_5Co[P(OCH_3)_3]_3$	Aromatic rings
$HCo(CN)_5^{3-}$	1,3-Dienes, $-CH{=}CHX$ where $X{=}COR$, COOH, CN, epoxides, alkyl halides
$NiCl_2/NaBH_4$	Alkenes,
(Arene) $Cr(CO)_3$	1,3-Dienes
$Na_2IrCl_6/(C_2H_5O)_3P$	Carbonyl group

* Aldehydes are decarbonylated by $RhCl(PPh_3)_3$.

References

1. W. Cocker and T. B. H. McMurry: *J. Chem. Soc.* 4549 (1956); W. Cocker, N. J. H. Dodds, and T. B. H. McMurry: *Tetrahedron* **3**, 160 (1958).
2. J. C. Banerji, D. H. R. Barton, and R. C. Cookson: *J. Chem. Soc.*, 5041 (1957).
3. J. B. Hendrickson and T. L. Bogard: *J. Chem. Soc.*, 1678 (1962).
4. A. E. Greene, J.-C. Muller, and G. Ourisson: *J. Org. Chem.* **39**, 186 (1974).
5. A. C. Day, J. Nabney, and A. I. Scott: *J. Chem. Soc.*, 4067 (1961).
6. A. J. Birch and K. A. M. Walker: *Australian J. Chem.*, 513 (1971).

7. M. Brown and L. W. Piszkiewicz: *J. Org. Chem.* **32**, 2013 (1967).
8. H. C. Odom and A. R. Pinder: *J. Chem. Soc. Perkin I Trans.*, 2193 (1972).
9. F. J. McQuillin and W. O. Ord: *J. Chem. Soc.*, 2902 (1959).
10. R. Howe and F. J. McQuillin: *J. Chem. Soc.*, 2670 (1956).
11. D. R. Fahey: *J. Org. Chem.* **38**, 80 (1973).
12. G. Wilke: *Angew. Chem.* **75**, 10 (1963); M. Ohno and M. Okamoto: *Tetrahedron Letts.*, 2423 (1964).
13. D. K. Fukushima and T. F. Gallagher: *J. Amer. Chem. Soc.* **77**, 139 (1955).
14. H. T. Brodie, C. E. Hay, and T. A. Willstruck: *J. Org. Chem.* **37**, 3361 (1972).
15. C. Djerassi and J. Gutzwiller: *J. Amer. Chem. Soc.* **88**, 4537 (1966); cf. H. J. Brodie, K. T. Kripalani, and C. Possanza: *ibid.* **91**, 1241 (1969).

INDEX OF SUBJECTS

GPSR Compliance

The European Union's (EU) General Product Safety Regulation (GPSR) is a set of rules that requires consumer products to be safe and our obligations to ensure this.

If you have any concerns about our products, you can contact us on ProductSafety@springernature.com

In case Publisher is established outside the EU, the EU authorized representative is:

Springer Nature Customer Service Center GmbH
Europaplatz 3
69115 Heidelberg, Germany

Batch number: 10607224

Printed by Printforce, the Netherlands